EXECUTIVE SUMMARY

The character of future international conflicts represents a complex and unpredictable set of challenges that necessitates a significant shift in the United States' approach to warfighting. Strategic guidance in *Sustaining U.S. Global Leadership: Priorities for 21st Century Defense* reinforces that *"...the United States will continue to take an active approach to countering...threats by monitoring the activities of non-state threats worldwide, working with allies and partners to establish control over ungoverned territories, and directly striking the most dangerous groups and individuals when necessary."* The *U.S. Air Force (USAF) Remotely Piloted Aircraft (RPA) Vector—Vision and Enabling Concepts: 2013–2038* balances the effects envisioned in the *USAF Unmanned Aircraft Systems Flight Plan 2009–2047* with the reality of constrained resources and ambitious national strategy for a complex world. More importantly, as a visionary document, the RPA Vector opens the aperture beyond current austere fiscal realities to explore art of the possible technologies in the 2013–2038 timeframe. The intent is to examine technological advances necessary to enable the Air Force's future RPA force.

The large-scale, complex, force-on-force scenarios that drove much of the military's "traditional" approach to planning are no longer sustainable. The Air Force will continue to plan in an effort to prepare for potential scenarios, whatever they may be. The Joint force will need to be smaller, leaner, adaptable, and make selective investments to meet its missions for United States national security at home and abroad, and the Air Force will play a significant role in this effort.

RPA and the effects they provide will continue to have an important role as we rebalance the force toward the Asia-Pacific region. The essential problem for future joint forces will be to project military force into an operational area and sustain it in the face of armed opposition. The Chairman of the Joint Chiefs of Staff (CJCS) *Joint Operational Access Concept (JOAC)* states that, *"Gaining friendly operational access can involve interdicting the enemy's force projection through the employment of one's own anti-access capabilities and geography, particularly distance, arguably determines the access challenge more than any other factor. ...To meet this challenge, future joint forces must leverage cross-domain synergy to establish superiority in some combination of domains that will provide the freedom of action required by the mission."* This is further supported by the *Defense Budget Priorities and Choices*, which directs that *"The resultant joint force must be multirole capable and able to handle our most demanding contingency plans, including Homeland Defense."*

The focus on the Asia-Pacific region places a renewed emphasis on air forces and cross-domain synergy. The unique challenge of the region reemphasizes the importance of understanding the Air Force's interconnected framework of vision, operating concepts and doctrine to meet the new security challenges.

> *"From airpower's earliest days, airmen have exploited technology to provide essential knowledge and information on when and where to act ... to control the ultimate high ground and to strike when and where directed.."*
>
> *– Air Force Posture Statement 2013*

The Next-Generation (NextGen) RPA must support the Secretary of Defense's (SECDEF) stated need to *"project power in contested and [anti-access/area denial] A2/AD environments, strike quickly from over the horizon..."* In addition, NextGen systems must further enable cross-domain synergies, which drive the need for a family of RPA and Small Unmanned Aircraft Systems (SUAS), ranging from micro- and nano-sized vehicles to medium/large NextGen RPA. This NextGen RPA must be multi-mission capable, adverse weather capable, net-centric, interoperable and must employ appropriate levels of

autonomy. In addition, potential future conflicts will drive the need for a long-range, long-endurance capability. The past *Unmanned Aircraft System (UAS) Flight Plan* articulated an unprecedented vision for RPA and ultimately reinforced the importance of vision statements and their impact on the Air Force's future as envisioned in the recent update to Air Force Doctrine Document (AFDD)-1.

> "...*To put this into perspective is to understand the different uses of vision, operating concepts, and doctrine. If placed along a continuum, doctrine, operating concepts, and vision provide a model for thinking about future technology, operating constructs, and doctrine in a coherent temporal framework.*"
>
> *– Air Force Doctrine Document-1*

Through this interconnected framework of doctrine, operating concepts and vision, it is possible to examine a wide range of doctrine, organization, training, materiel, leadership and education, personnel and facilities (DOTMLPF) issues associated with RPA. We must continue to build upon our success in the development of current RPA, sensors, operating concepts and infrastructure. The past tendency of rapidly introducing new and untested systems has allowed the quick reaction fielding of capability to support warfighter need. Though necessary to meet growing warfighter demands, this has also resulted in gaps in sustainment capability, proprietary stove-piped systems, production delays and excessive life-cycle cost growth. The future will require deliberate acquisition of systems that are open, modular, and rapidly adaptable to the broad range of military operations. RPA continue to represent a wide range of DOTMLPF challenges that must be addressed at the appropriate point during technology development, concept exploration, and systems acquisition.

This document, *RPA Vector: Vision and Enabling Concepts (2013–2038),* is intended to replace the *Air Force UAS Flight Plan* 2009–2047, which pre-dated the Core Function Lead Integrator (CFLI) construct. It is intended as a strategic planning document to inform the capabilities planning and requirements development process across multiple Service Core Function portfolios. The CFLI, when executing responsibilities in the strategic planning process, should reference this document to link appropriate Core Function Master Plans (CFMP). This document will be reviewed and updated by Headquarters Air Force within 2 years in coordination with CFLI staffs. The RPA Vector is intended as a reference to guide development of Air Force CFMPs. Whether or not to pursue RPA as the solution for a specific requirement is at the discretion of the respective CFLI and must take into account interoperability of data systems, data links, interfaces, waveforms, weaponry, architecture standards and airspace access constraints. The RPA Vector considers the long-term impacts of advanced RPA technologies and concepts and describes key future operating environments and targeted operational capabilities to better focus technology investments. The development of future concepts necessitates further examination and validation of supporting doctrine for those emerging capabilities. The RPA Vector can provide a basis for future wargaming—the results of which may point to doctrinal considerations requiring further examination. It also underscores the need for increased RPA capability development and synchronization within the Air Force's various budgeting, programming, and requirements processes. Increased synchronization will ensure a more balanced weapons system portfolio, as the Air Force continues to manage the anticipated resource decline associated with the future fiscal environment.

Deborah Lee James
Secretary of the Air Force

Mark A. Welsh III
General, USAF
Chief of Staff

Table of Contents

LIST OF TABLES

LIST OF FIGURES

> *"The past decade for RPA mirrors the rapid evolution of combat airpower during World War I: a wave of great ideas, tactics, and technology, brought from air-minded communities flowed in faster than our ability to field them and slower than the land forces would have liked them. But like the Rickenbackers and Lufberys of their day, it was the RPA lieutenants and captains, staff sergeants, and senior airmen who took these new instruments of airpower, as imperfect as they were, and integrated them into the evolving fight, transitioning the platforms from reconnaissance-only to true multirole Intelligence, Surveillance, and Reconnaissance (ISR) and strike. They delivered disciplined and effective combat airpower every day; another generation of the Air Force's great captains is born."*
>
> *– RPA Expeditionary Operations Group Commander (2010-2012), Colonel Bill Tart*

1.   INTRODUCTION

Over the last decade, remotely piloted aircraft (RPA) have become a critical component in the application of airpower and one of the most "in demand" platforms the Air Force provides to the joint force. RPA accentuate the core tenets of persistence, flexibility, and versatility of mission. Along with information collection and attack capabilities, they have a proven record as force multipliers during the contingency operations of the last decade. Their utility across the range of military operations (ROMO) remains mostly untested, as RPA, similar to all new acquisitions in the last decade, have not routinely operated in contested airspace or in an A2/AD environment. Current systems have operated in contested threat environments over Baghdad and Libya, though operations were dependent upon the type of platform and network capability. Today, RPA provide near-real-time information, not only to senior operational decision makers but also directly to joint and coalition forces in the field. RPA aid forces in combat currently perform selective precision strike missions against pre-planned or high-value opportunities, minimizing risk of collateral damage.

> *"We already combine our air, space, and cyber forces to maximize these enduring contributions, but the way we execute must continually evolve as we strive to increase our asymmetric advantage…Our airmen's ability to rethink the battle while incorporating new technologies will improve the varied ways our Air Force accomplishes its missions."*
>
> *– Chief of Staff United States Air Force, General Mark Welsh*

The Air Force has taken preliminary steps toward a comprehensive service-wide review of future RPA applications. In addressing the continued growth of unmanned aircraft systems (UAS), the Air Force will consider a broad doctrinal look at the kind of missions RPA have supported in Operation IRAQI FREEDOM (OIF) and Operation ENDURING FREEDOM (OEF) and will look at the strengths and challenges as well as the current and future demands of the combatant commanders (CCDR) to understand how all components and capabilities can contribute to the successful execution of all phases of conflict.

> *"The United States faces profound challenges that require strong, agile, and capable military forces whose actions are harmonized with other elements of U.S. national power."*
>
> *– Sustaining U.S. Global Leadership: Priorities for 21st Century Defense*

This level of understanding is needed to determine what forces and resources need to be established in baseline budgets from which we would surge in crisis to achieve operational plans strategies, what risk we are willing to accept, or how such risk could be mitigated.

1.1   Purpose And Scope

Both the current fiscal and future operational environments facing the Air Force influence the landscape for investments in the development and fielding of new technologies.   This document refines the Air Force strategic vision for the future of RPA and reemphasizes the inherent potential and emerging capabilities of small unmanned aircraft systems (SUAS).   The RPA Vector outlines concepts and capabilities needed over the next 25 years.   It can inform the capabilities planning and requirements development process as well as inform the CFLIs as they execute their responsibilities for implementation planning in the plans, programming, budgeting and execution process.

Technology advancements must increasingly leverage open architecture, autonomy, modularity, and interoperability to maximize both system effectiveness and service investments. Open architecture and open interfaces need to be leveraged to address problems with proprietary system architectures. Standards and interface specifications need to be established to achieve modularity, commonality, and interchangeability across payloads, control systems, video/audio interfaces, data, and communication links.   This openness will enhance competition, lower life-cycle costs, and provide warfighters with enhanced unmanned capabilities that enable commonality and joint interoperability on the battlefield. Focused investments must be made in today's fiscally constrained environment to further the Air Force's strategic, operational, and tactical capabilities to support joint operations.   These joint forces will possess the capabilities to deploy, operate, and employ autonomous behaviors to reduce pilot and operator workload.   The future Air Force RPA force structure will be leaner and more adaptable to maximize the effectiveness of 21st-century airpower.   The Air Force will continue to be organized for operations to support strategic missions.

1.2   Vision

The Air Force vision for RPA is—

- Seamless integration of RPA into operations across all domains (air, space and cyberspace) and levels of warfare (strategic, operational and tactical) will enable future joint forces to leverage cross-domain synergies in environments ranging from permissive to non-permissive.
- Widespread use of highly adaptable and flexible autonomous systems and processes to provide significant time efficiencies and operational advantages over adversaries.
- Increasingly interoperable systems and platforms through advancements in and application of open architecture, standards, and modularity across joint, interagency and coalition partners that results in a leaner, more effective, adaptable, sustainable and efficient force.
- Teaming across departments and agencies, coalition partners, academia, and industry to drive innovation, technology, and efficient use of research and development (R&D) investments.

The family of systems (FoS) and capabilities outlined in this document reflect current funded programs, Defense Planning Guidance, Annual Planning and Programming Guidance, and Core Function Master Plan (CFMP) guidance, science and technology (S&T) areas of research and proposed future concepts.

The rapid development, growth, and integration of RPA to provide support to ongoing irregular warfare (IW) contingencies; direct support for non–Department of Defense (DoD) government agencies; support to humanitarian/disaster relief operations; and urgent support to other federal and civil efforts have overshadowed efforts to determine the characteristics of future operating environments and the suitability of emerging technologies.

1.3   Assumptions

Eight key assumptions guided the development of the RPA Vector:

- Concept of Operations (CONOPS) and capabilities that directly support the National Defense Strategy will have the highest priority.
- Manned and remotely piloted systems will be seamlessly integrated and synchronized in airspace, command and control (C2) architectures, information sharing, and purpose to optimize airpower and increase capability across the full ROMO for the combined force.
- Manned and remotely piloted systems will require cross-domain information exchange to provide the freedom of action required by the mission.
- Capabilities must be focused to enable unconstrained operational access when a common air, space, or cyberspace domain is unusable or inaccessible.
- Secure, efficient, robust, agile and redundant C2 and information distribution will be essential for mission success in all environments.
- Increased use of autonomy and autonomous behaviors will increase capability in permissive and non-permissive environments while leveraging operating cost savings.
- Open architecture and standardized interfaces will enable modularity, resulting in increased warfighter effect, adaptability, sustainability, interoperability and reduced life-cycle cost.
- Impacts of budget sequestration can be mitigated.

1.4   Systems Descriptions and Terminology

UAS consist of a control station, one or more unmanned aircraft (UA), control and payload data links, and mission payloads designed or modified not to carry a human pilot and operated through remote or self-contained autonomous control. A UA is defined in Joint Publication 1-02 as *"an aircraft or balloon that does not carry a human operator and is capable of flight under remote control or autonomous programming."* Within the DoD, these aircraft have been categorized by weight, operating altitude, and airspeed, as delineated in Table 1. The Air Force has further defined Group 1 through 3 as SUAS and Group 4 and 5 as RPA.

1.4.1   Small Unmanned Aircraft Systems

An SUAS is a Group 1, 2, or 3 UA. Examples of Air Force SUAS include the RQ-11 Raven and Wasp III. SUAS can be operated remotely or on pre-programmed autonomous routes, can be expendable or recoverable, and can carry a lethal or nonlethal payload. SUAS operators (SUAS-O) are not rated pilots but function as the pilots in command (PIC) and are responsible for the safe ground and flight operation of the UA and onboard systems. Air Force SUAS-Os are selected, trained, and certified to act as the PICs of their UA based on mission requirements. SUAS-O qualification does not result in award of an Air Force Specialty Code (AFSC) or a Special Experience Identifier.

1.4.2   Remotely Piloted Aircraft

Group 4 and 5 UAS are classified as RPA. Examples of RPA include the MQ-1 Predator (see Figure 1), MQ-9 Reaper, RQ-4 Global Hawk, and RQ-170 Sentinel. An RPA requires a rated pilot (11U or 18X AFSC), sensors operator or system operator, a ground control station (GCS) and squadron operations center (SOC), associated manpower and support systems, and communication infrastructure to perform mission and intelligence integration. The RPA design is unique in that it is not constrained by life support elements and size or weight of the person operating the aircraft. RPA provide the intrinsic benefits from combinations of persistence, endurance and maneuverability. Given that they are remotely piloted, potential RPA operational environments can include permissive, contested and A2/AD without exposing the aircrew to risks. This does not imply all RPA are expendable—on the contrary, many of these aircraft are specialized and expensive commodities. Therefore, commanders will have to make very careful risk-benefit assessments when employing RPA assets in contested and A2/AD environments.

Table 1: Representative Air Force Platforms versus Joint UAS Group Classification

	UAS Groups	Maximum Weight (lbs)	Normal Operating Altitude (ft)	Speed (kts)	Representative Aircraft
SUAS	Group 1	0-20	< 1,200 AGL	<100	Raven (RQ-11) / Wasp
SUAS	Group 2	21-55	< 3,500 AGL	< 250	Scan Eagle
SUAS	Group 3	< 1,320	< FL 180	< 250	Currently No USAF Program of Record for this Category
RPA	Group 4	> 1,320	< FL 180	Any Airspeed	Predator (MQ-1)
RPA	Group 5	> 1,320	> FL 180	Any Airspeed	Reaper (MQ-9) / Global Hawk (RQ-4)

RPA engage in many of the same missions as manned aircraft, such as close air support (CAS), ISR, dynamic targeting, and air interdiction, airborne interdiction of maritime targets, maritime air support, strike coordination and reconnaissance (SCAR), communications relay, and combat search and rescue (CSAR). In most cases, future RPA will require access to an interoperable, affordable, responsive, and sustainable networked system of systems (SoS) capable of satisfying service, joint, interagency and coalition tactical information exchanges. This system must be distributed, scalable and secure. It includes but is not limited to human interfaces, software applications and interfaces, network transport, network services, information services, and the hardware and interfaces necessary to form a complete system that delivers tactical mission outcomes. The network operates as independent, small combat sub-networks connected to each other and to the Global Information Grid (GIG). The advantages of this structure make worldwide real-time information available to the aircrew as well as worldwide real-time dissemination of information from the RPA to the tactical edge. Terrestrial-based resources and connectivity allow additional tools and intelligence analyst resources to be called upon on demand when and where needed. In other scenarios and operating environments, particularly those with limited or no access to the GIG, systems must retain the flexibility to deploy and operate via available connectivity solutions, including line-of-sight (LOS) communications.

Figure 1: MQ-1 Predator Returning from Mission

2.   CURRENT OPERATIONS

Current DoD strategy highlights that the next 10 years will require renewed focus on solving challenges confronting a fundamental American military mission—global power projection. Time and resources must be invested now to organize, train, and equip the force to sustain American projection of global power (see Figure 2).

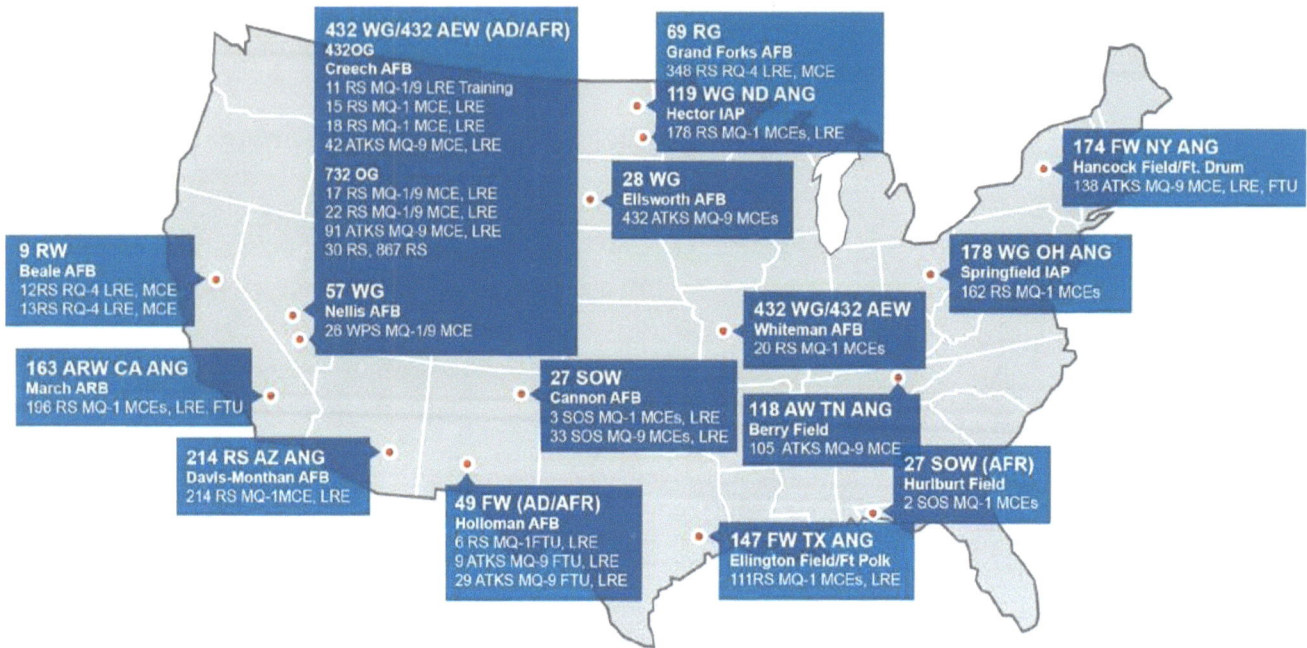

Figure 2: Current RPA Operating Locations

This guidance requires the Air Force to reassess how to effectively and efficiently posture itself for the future security environment. The *JOAC* describes how the future *"joint force will leverage cross-domain synergy—the complementary vice merely additive employment of capabilities in different domains such that each enhances the effectiveness and compensates for the vulnerabilities of the others—to establish superiority in some combination of domains that will provide the freedom of action required by the mission."*

> *"As we end today's wars and reshape our armed forces, we will ensure that our military is agile, flexible, and ready for the full range of contingencies. In particular, we will continue to invest in the capabilities critical to future success, including ISR...operating in anti-access environments and prevailing in all domains, including cyber."*
>
> *– Sustaining U.S. Global Leadership:*
> *Priorities for 21st Century Defense*

2.1   RPA Force Structure

The primary RPA force structure consists of MQ-1/9 and RQ-4. The current active duty MQ-1/9 squadron construct provides five combat air patrols (CAP) and the manpower to support aircrew, maintenance, communications, weapons loaders and security. Air National Guard (ANG) MQ-1/9 squadron construct provides up to three CAPs. The Air Force continues to fully institutionalize RPA by committing the manpower and fiscal resources to train and equip a robust capability in MQ-1/9 and other platforms (see Figure 3).

The RQ-4 Global Hawk fleet consists of Block 20, Block 30, and Block 40 aircraft. The Block 20 system provides communications relay with the Battlefield Airborne Communications Node (BACN) payload. The Block 30 system provides a multi-intelligence collection capability through a combination of electro-optical/infrared (EO/IR) and synthetic aperture radar (SAR) imagery and a signals intelligence (SIGINT) payload. The Block 40 system provides long-range, high-resolution SAR imagery and ground moving target indicator (GMTI) capability. The current fleet provides aircraft to sustain three geographically separated 24x7 orbits worldwide.

> *"[The U.S.] defense strategy...advances the Department's efforts to rebalance and reform, and it supports the national security imperative of deficit reduction through a lower level of defense spending."*
>
> *– Sustaining U.S. Global Leadership: Priorities for 21st-Century Defense*

By 2020, the Air Force expects to have the smallest force structure in its history. Balancing future capability needs, significant decreases in future funding, and the transition of current operational capabilities from the Overseas Contingency Operations (OCO) funds to the base budget will require a disciplined, focused, and centralized RPA strategy.

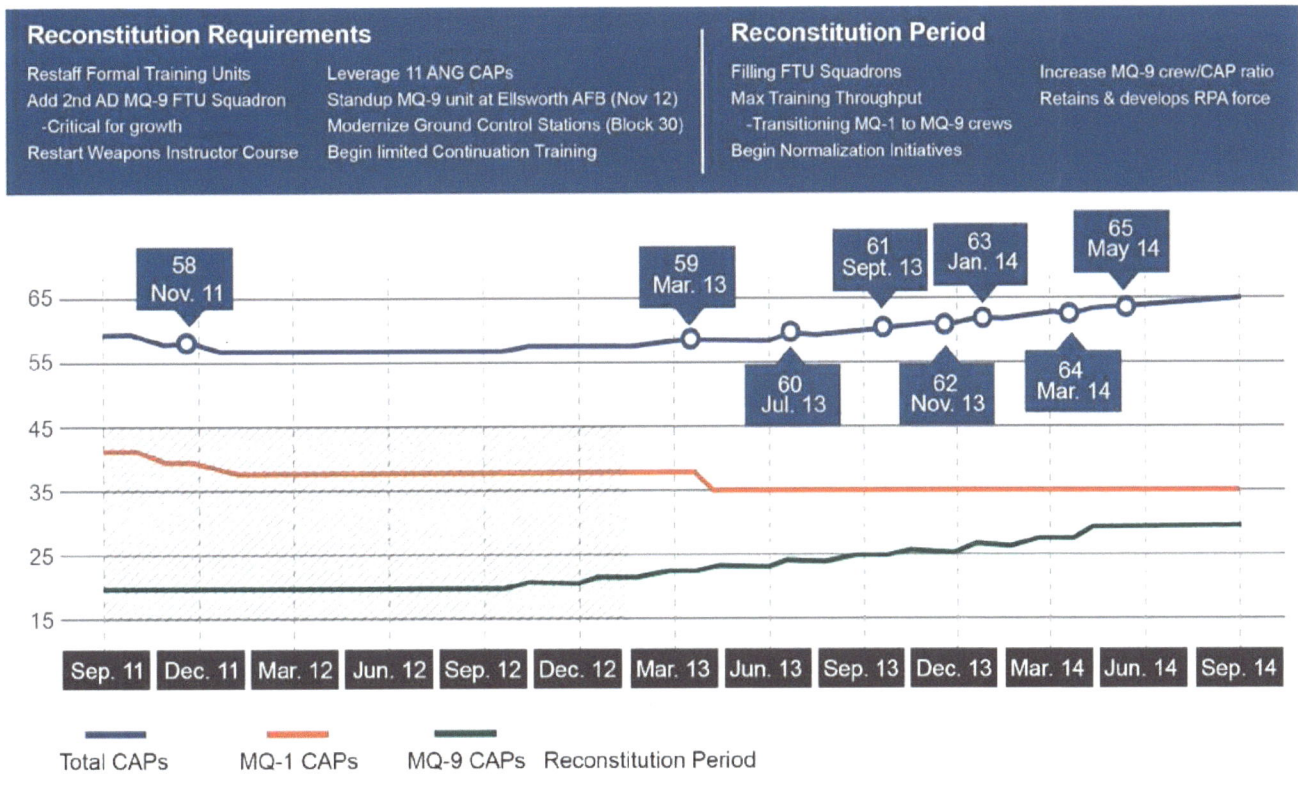

Figure 3: MQ-1/9 CAP Growth

The Air Force RPA force structure completed a reconstitution period to ensure operational sustainment following an operational surge and surpassing 2 million combat hours in October 2013. The reconstitution enables a 65 MQ-1/9 CAP capability by May 2014. Steps to reconstitute the RPA fleet included—

- Remotely piloted fleet steady state consisting of just less than 400 aircraft
- 24 MQ-1/9 units at 18 continental United States (CONUS) locations (including 11 ANG units)

- RPA personnel cadre and experience (aircrew, intelligence, aircraft maintenance, communications maintenance) growth from 2,100 personnel (2005) to 9,900 (2017).

Support to critical missions over the last 10 years has driven numerous surges in RPA CAP growth. This necessary growth has not come without a cost; the surges have required the Air Force to stand down critical training capacity and reprioritize RPA equipment away from other vital functions, such as test and training. During fiscal year (FY) 2012, more aircrew went through initial qualification training for RPA than for fighters or bombers combined.

By May 2014, the Air Force will provide 65 MQ-1/9 CAPs, per 29 June 2011 SECDEF memo, with the capacity to surge additional CAP capability for emerging contingency situations. In May 2014, the 65 CAPs will consist of approximately 33 MQ-1 and approximately 32 MQ-9s, and the Air Force will begin to transition to an all MQ-9 fleet (see Figure 4).

Figure 4: MQ-9 Firing an AGM-114 Hellfire

2.2 RPA Operations

RPA have been employed to conduct CAS, ISR, CSAR, air interdiction and SCAR missions. In the future, RPA could be considered for a broader range of uses, such as cargo, aerial refueling, and others. Many RPA and their aircrews provide critical combat support functions, such as vehicle follows and special operations support as well as responses to time-critical events like troops-in-contact, while others are equipped with specialized sensors for collection of imagery and other intelligence information (see Figure 5). RPA integrate with air and ground forces from all other services and coalition partners.

Figure 5: MQ-9 Ground Control Station

Over the battlefield, MQ-1/9 RPA aircrews are capable of communicating in real time with joint terminal attack controllers (JTAC), ground forces, supported units, air traffic control (ATC) and other air assets. During an actual engagement, RPA crews communicate directly with the JTAC via secure radio onboard the aircraft or through a number of other alternate means. Many RPA are capable of providing real-time full-motion video (FMV) to the JTAC, the tactical operations center (TOC), the ground forces commander, or any other ground forces through remotely operated video-enhanced receiver (ROVER) or other remote video terminals (RVT). ROVER has been effective in more than 1,000 weapons employments, and additional capabilities continue to be added to future generations of ROVER as new technologies become available. In the future, emerging technologies will allow ground forces to provide enhanced targeting information back to RPA via NextGen RVT.

The Global Hawk (see Figure 6) Block 40 Multiplatform Radar Technology Insertion Program (MP-RTIP) radar is a side-looking active electronically scanned array (AESA) multimode radar, featuring a variety of improved GMTI surveillance modes and high-resolution SAR capabilities to detect, track, and support combat identification (CID) of ground targets. The MP-RTIP sensor can collect and process GMTI and SAR simultaneously. AESA technology provides the beam agility to rapidly interleave radar modes. Software internal to the radar manages modes and the antenna beam using a priority scheme that takes into account both operator-established priorities and internal requirements for successful mode operation. As net-centric operations evolve in the future, the MP-RTIP sensor will be well-suited to accommodate multiple mission objectives in support of intelligence exploitation organizations, ground units, and battle management nodes. Future upgrades will provide enhanced capabilities to detect and support CID of ground, air and maritime targets.

Figure 6: RQ-4 Global Hawk

RPA integration with the operations centers, Air Force Distributed Common Ground System (DCGS), and theater C2 nodes is critical to mission success. The unique characteristics of RPA and remote split operations (RSO) offer challenges that must be overcome to combat the geographic separation of these entities. As an example, a recent improvement in overcoming one of these challenges is that Distributed Ground Station (DGS) personnel now have the ability to interact and verbally communicate directly with the RPA aircrew, which was previously only available via text-based chat.

2.3   Remote Split Operations

RSO refers to the geographical separation of the launch and recovery cockpit and crew from the mission cockpit and crew. RSO enables the employment of the aircraft by the mission crew at a location other than where the aircraft are based (in some cases, thousands of miles from the actual aircraft location). RPA using the RSO method of employment increases the percentage of assets available for operations due to the distributive nature of RSO. The resulting deployment and employment efficiencies lend greater flexibility and may provide capability at the same or reduced expense when compared to manned equivalents. Further, with the reduced manpower footprint required forward, there are fewer force protection and logistics support requirements to support operations. Figure 7 depicts the current RSO architecture.

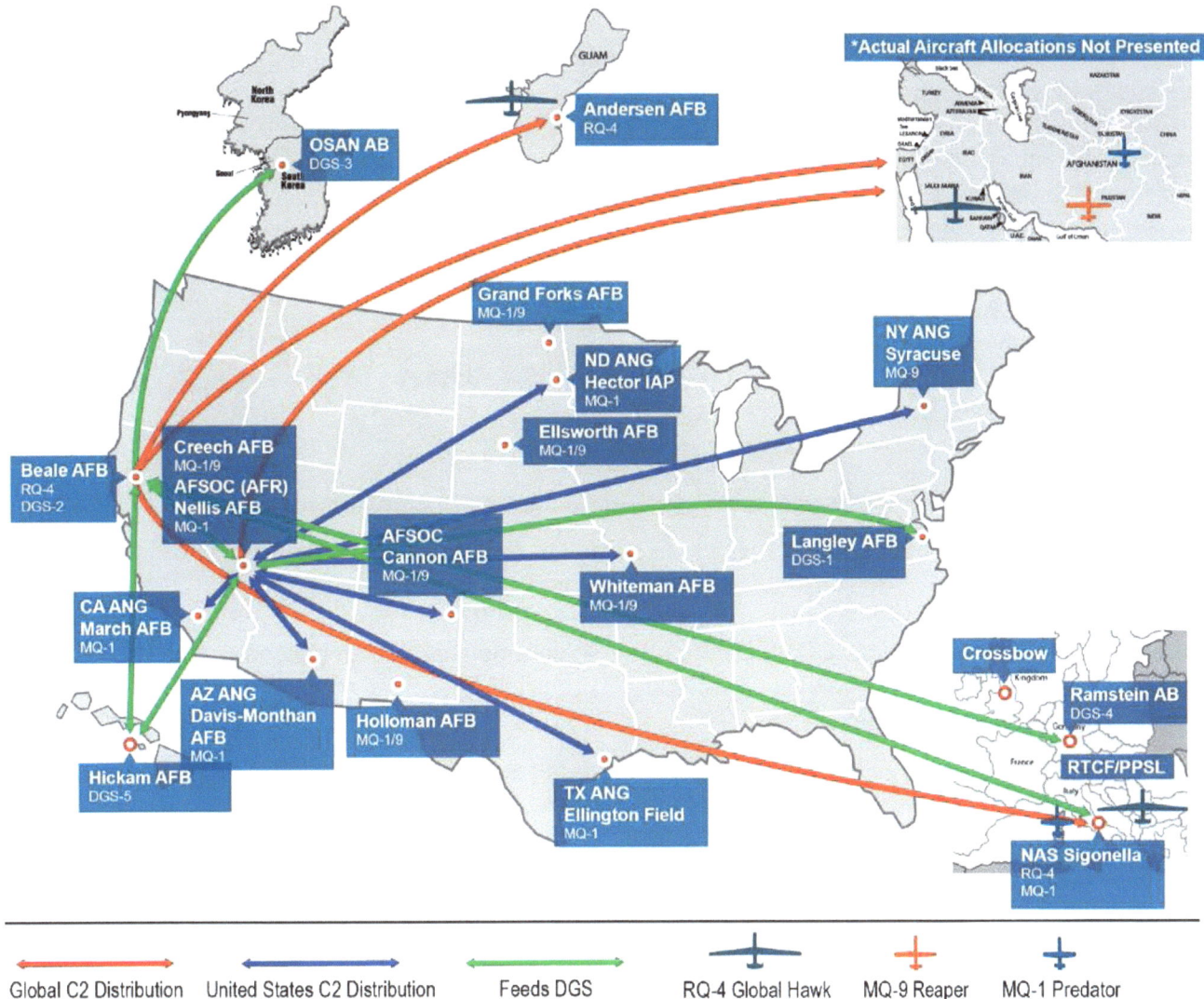

Figure 7: RSO Architecture

RPA provide a new level of global agility by maximizing the advantages of the inherent capabilities associated with the RSO concept. The foundation to the RSO concept is the communications architecture. However, the current RPA global communications architecture is not included under an existing program of record (PoR) and is only funded via OCO or available operations and maintenance funds; this lack of PoR funding places the critical RPA global communications architecture on unsure footing.

Air Force Space Command's (AFSPC) recent RPA global communications architecture review highlighted the lack of Air Force entity programming for RPA communications infrastructure as one of the two key RPA architecture problem statements. On 1 August 2012, Secretary of the Air Force designated the cyberspace superiority CFLI for the responsibility to manage RPA global communications infrastructure.

The current RPA communications architectures were designed and fielded leveraging all communications domains to extend RPA C2 and data beyond line-of-sight (BLOS) of the RPA's operating area. This architecture established a new paradigm of globally distributed C2. However, due to the rapid fielding of RPA systems to support geographic combatant command (GCC) requirements, the RPA communications architecture was neither optimized nor was a future architecture fully envisioned. Today's RPA rely on communications for C2 and dissemination of information.

As RPA crews participate in complex mission sets that include precision weapons effects (e.g., CAS, SCAR, personnel recovery [PR]), C2 communications become increasingly important. Diverse target sets and unknown conditions stress C2 communication and data links beyond today's capabilities. The Air Force must continue to address frequency and bandwidth availability, link security, link ranges, assuredness, reliability and network infrastructure to ensure continuous undisrupted availability for operational and mission support of remotely piloted systems.

> *"Modern armed forces cannot conduct high-tempo, effective operations without reliable information and communications networks and assured access to cyberspace and space. Today space systems and their supporting infrastructure face a range of threats that may degrade, disrupt, or destroy assets. Accordingly, DoD will continue to work with domestic and international allies and partners and invest in advanced capabilities to defend its networks, operational capability, and resiliency in cyberspace and space."*
>
> *– Sustaining U.S. Global Leadership: Priorities for 21st-Century Defense*

The global reach capability of RSO provides the means to balance and surge efforts across theaters and to serve multiple CCDRs from a unified RPA command structure. The supporting communications architecture is complex and one of the major cost drivers for RPA. Currently, AFSPC is designated as the lead Major Command (MAJCOM) for RPA global communications architectures, and the Assistant Secretary of the Air Force for Acquisition (SAF/AQ) selected Space and Missile Systems Center as the responsible system program office (SPO).

There are several issues facing Air Force efforts to develop an affordable, interoperable RPA communication infrastructure that can support operations across all phases of conflict. First, there is no centralized communications enterprise governance. In addition, there is no single entity responsible for the management and programming across the entire RPA communications infrastructure. AFSPC as the lead command will have the responsibility to program for and manage the entire RPA communications enterprise infrastructure. In addition, the Secretary of the Air Force Office of Information Dominance and Chief Information Officer, (SAF/CIO A6) has been given overarching governance authority of the RPA global communications architecture, aligned under the Warfighting Integration General Officer Steering Group.

2.4   Globally Integrated RPA Data Infrastructure

The rapid growth in RPA has created significant challenges for the Air Force. Data utilization (handling and storage) and manpower are critical elements for the execution and exploitation of the full range of capability provided by RPA. These capabilities include geospatial intelligence (GEOINT), SIGINT and communications.

> *"The ability to create precise, desirable effects with a smaller force and a lighter logistical footprint depends on a robust ISR architecture. Across all domains, we will improve sharing, processing, analysis, and dissemination of information to better support decision makers."*
>
> *– The National Military Strategy of the United States of America*

As the number of RPA CAPs grows, the sensor data is estimated to increase by more than 5,000 percent. The existing infrastructure used to fuse RPA data is insufficient to meet the projected increased demand. Improved collaboration is required between RPA operations centers and DCGS via the DCGS Integration Backbone (DIB). The DIB is a cohesive set of modular, community-governed, standards-based data services focused on enterprise information sharing. DIB provides a common framework for data exposure and transformation and for enabling applications and users to discover and access information from a wide range of distributed sources. This information access and collaboration will allow operations centers to complement the exploitation efforts through mission-driven resource management. Figure 8 depicts the current DCGS network.

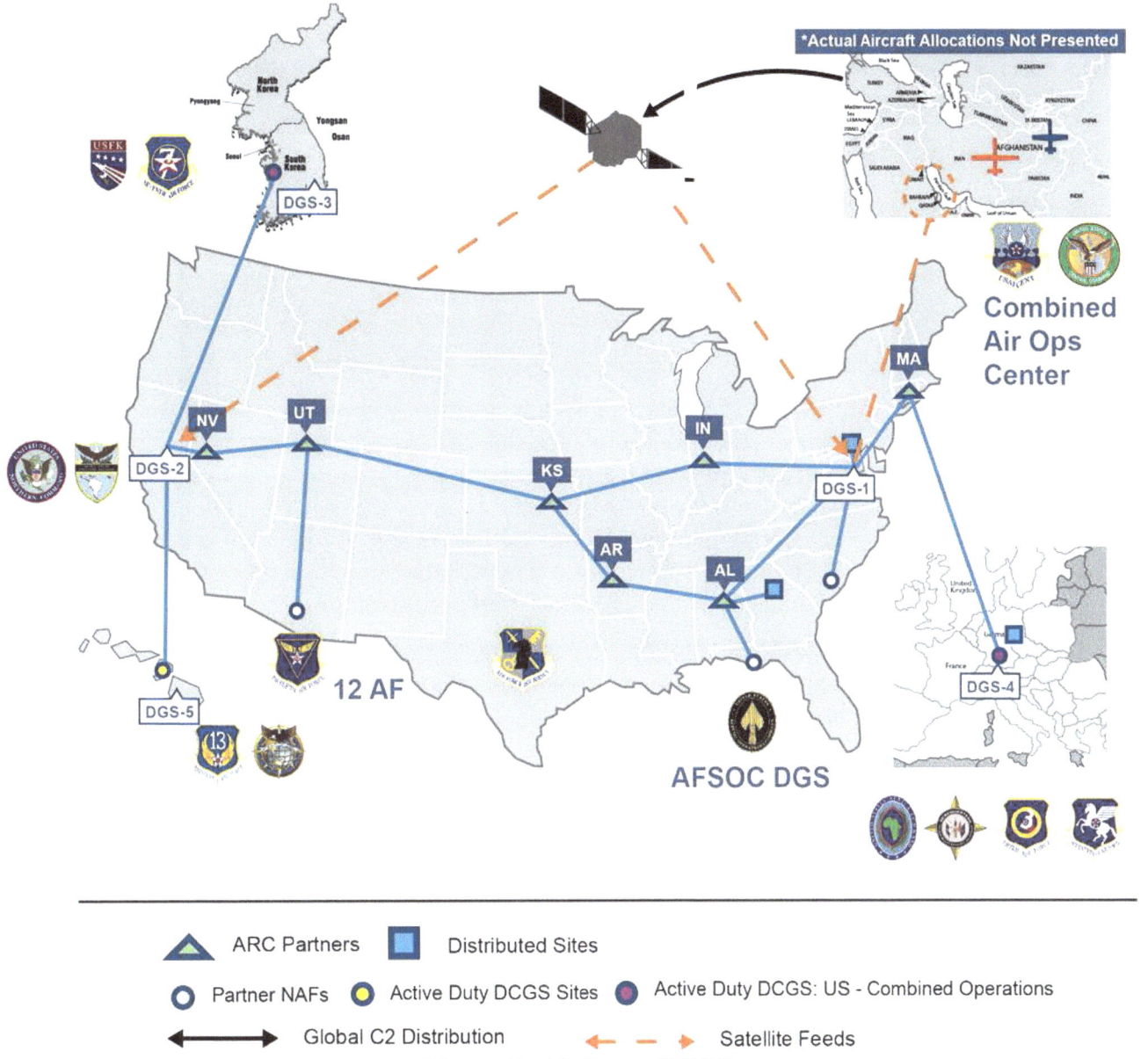

Figure 8: Air Force DCGS

2.4.1   Exploitation Needs

The Air Force DCGS supports ISR operations for both manned and remotely piloted platforms. The Air Force DCGS weapon system produces intelligence information collected by the U-2 Dragonlady, RQ-4 Global Hawk, MQ-9 Reaper, and MQ-1 Predator and provides key capabilities for intelligence fusion. Air Force DCGS is composed of geographically separated, networked sites. The individual nodes are regionally focused and paired with their corresponding Air Force component-numbered air force to provide critical processing, analysis and dissemination of ISR data collected within the numbered air force's area of responsibility (AOR). Collaboration between RPA and DCGS will remain critical and must continue to occur to ensure mission success.

The current Processing, Exploitation, Dissemination (PED) process for data exploitation is manpower-intensive, involving both DCGS and squadron-level intelligence analysis. Dynamic operations dictate the increased reliance on squadron-level intelligence personnel to provide time-sensitive support to RPA aircrews. DCGS crews are geographically separated from RPA crews.

Current RPA squadrons and DCGS have been built based upon models of existing manned aircraft organizations. In addition, ANG and Air Force Reserve (AFR) components are an integral part of Total Force Integration and provide an essential element to RPA operations.

As RPA operations continue to expand, the estimated manpower needed to support future RPA operations (including operations centers and DCGS) is estimated to exceed 9,900 personnel by FY2017. Continued increases in the number of intelligence personnel (in both RPA squadrons and PED squadrons) needed for each additional RPA CAP has led to personnel and process efficiency efforts to minimize manpower costs.

2.4.2 Information Synchronization

Reachback offers RPA organizations some unique advantages compared to most manned aircraft squadrons because RPA intelligence personnel are capable of providing the aircrew with real-time intelligence updates and products. RPA intelligence personnel are co-located with the RPA mission crew. The Mission Intelligence Coordinator (MIC) and Senior MIC are co-located with the RPA aircrew and are part of the RPA mission crew. In many RPA squadrons, the MIC is physically in the GCS along with the pilot and Sensor Operator. In cases where the mission is being flown by a mobile GCS, the MIC cannot be co-located in the GCS, so support functions are conducted from within the SOC.

Real-time intelligence information can contribute to mission effectiveness by capturing information that may have otherwise not been available due to not having a means of conferring near-real-time or real-time intelligence updates to and from the aircrew. In addition, intelligence personnel with focused skills are dedicated to supporting the pilot with data collection, analysis and targeting.

To take advantage of this opportunity, Global Hawk and Predator operators in the field quickly identified the need and created operations centers (Global Hawk Operations Center and Predator/Reaper SOC) out of necessity. As operations centers become part of the RPA multivehicle C2 PoR, current and future hardware and software programs must be integrated for ease of access to intelligence and operational sources. This program will develop NextGen C2 applications that will be integrated into RPA operations. Once integrated, the RPA mission crews (pilot, sensor operator, MIC) will have streams of data to manage and relay to other RPA crews, manned platforms, supported units and mission partners to enhance mission effectiveness and situational awareness (SA).

A key near-term issue is—

- RPA operations centers are not part of a PoR and have been funded at the unit level. Aggregated costs for these operations centers are approximately $30 million. RPA operations centers will become a part of the RPA multivehicle C2 PoR starting in FY14.

As a critical element of RPA operations, the SOC must handle various types of data, facilitate data storage and retrieval, and enable clear and assured communication between organizations and platforms with information applicable to mission execution. Information is exchanged between the aircraft, operations centers, and DCGS as well as between each service DCGS via the DIB. There are, however, significant differences between Air Force and the other services' operations. Air Force RPA typically operate BLOS, and operations centers leverage GIG connectivity. In contrast, many Army, Navy and Marine assets operate primarily in LOS and with limited connectivity to the GIG and via local command and control, communications, and computer rather than RSO. An exception to this is Army Special Operations Forces (SOF) MQ-1C, which are now capable of reachback PED by Distributed Ground Station-SOF.

> *"...we will continue migration to a service-oriented architecture to handle the increasing quantities of ISR data that is integrated and delivered from emerging sensors and platforms operating in all domains. We will also improve our ability to move information securely and reliably over information pathways."*
>
> *– FY2013 Air Force Posture Statement*
> *Michael Donley, Secretary of the Air Force*

Operations centers have provided significant benefit to intelligence information collaboration. The ability to have relevant intelligence and other pertinent information readily available while collaborating within a User Defined Operating Picture (UDOP) provides operators and analysts the ability to focus on their specific mission rather than being overwhelmed by data. Operations centers give RPA the ability to integrate into the intelligence community through DCGS. Intelligence integration will continue to be addressed for Air Force DCGS and the joint community.

2.5   SUAS Operations

SUAS have had a profound effect on the battlefield by providing timely and assured information to tactical ground units. In conventional and special operations, Air Force JTAC, special operations weather teams (SOWT), security forces, and Office of Special Investigations agents have employed SUAS to provide point, route and environmental reconnaissance, target development, post-attack assessment of effects and other innovative applications. The Air Force recognized the unique utility and capabilities of SUAS during initial phases of OIF when the Air Force purchased Pointer SUAS for combat control units.

Furthermore, the SUAS FoS represents a unique approach and challenge to the larger manpower structures supporting UAS operations. Ultimately, SUAS airpower can be carried in a backpack with commensurate capabilities. SUAS are highly effective in supporting integrated, manned and remotely piloted mission sets beyond those met by the MQ-1/9 and RQ-4. Currently, SUAS serve as organic ISR/targeting assets for JTAC and SOWT. SUAS provide the commander and individual service members' life-saving SA and thus represent a profound technological advance in air warfare. The need for SA and GEOINT dominates current urgent requests from the field. SUAS have been successful in operations for targeting and strike. RPA and manned ISR aircraft are used to provide over-watch and strike capability when available to support high-priority task force missions with ground teams receiving FMV feed via RVT.

Characteristics unique to most SUAS include the use of basic encryption techniques, reliance on LOS data links, and isolation from existing ISR and PED architectures. Air Force and United States Special Operations Command (USSOCOM) invested roughly $110 million per year in OCO funding for contractor-operated Group 2 UAS for theater ISR support; those systems flew more than 6,700 hours in FY12 alone.

Figure 9: RQ-11 Raven Prior to a Mission

Air Force Special Operations Command (AFSOC) as lead command for the Air Force for SUAS, developed a vision that includes a requirement strategy encapsulated in the Air Force requirement oversight council (AFROC)-approved SUAS Initial Capabilities Document (ICD) (Ref: AFROC memorandum [AFROCM] 02-04-02 and 03-05-04) for an FoS approach. Within the strategy, the Air Force teamed with the Marine Corps and the Navy to develop the Tier II Small Tactical Unmanned Aircraft System (STUAS) ICD (Ref: joint requirements oversight council memorandum [JROCM] 021-07 and Capability Development Document AFROCM 08-08-03 and JROCM 219-08) for potential future Group 2/3 capability to support a wide variety of mission sets. A joint Analysis of Alternatives (AoA) was conducted for the Tier II STUAS to analyze future Air Force needs. SOF SUAS operational needs (see Figure 10) consist of—

- BLOS tactical ISR and targeting
- Kinetic low-collateral damage engagement of time-sensitive targets
- Rapid reaction expeditionary persistent ISR
- Near-real-time networked collaborative information
- Standoff, adverse weather capable, multiple target track/kill from AFSOC aircraft.

Figure 10: SUAS Operations

2.6   Airspace Integration

The number of RPA and SUAS platforms in the Air Force inventory and their corresponding roles are growing at a rapid pace. RPA training requirements have eclipsed the current special use airspace available and the projected demand will continue to increase. Although strides have been made integrating RPA into joint airspace abroad, the services must continue to pursue materiel and non-materiel solutions to better integrate UA into the National Airspace System (NAS).

> "DoD UAS require routine NAS access in order to execute operational, training, and support missions and to support broader military and civil demands. UA will not achieve their full potential military utility to do what manned aircraft do unless they can go where manned aircraft go with the same freedom of navigation, responsiveness, and flexibility."
>
> *– Unmanned Systems Integrated Roadmap FY2011–2036*
> *Office of the Secretary of Defense*

The Air Force, National Guard, Customs and Border Protection, and other interested governmental agencies must continue to work with the Federal Aviation Administration (FAA) to ensure routine UAS access to the NAS to meet training (see Figure 11) and operational needs supporting Defense Support of Civil Authorities (DSCA) requirements. The provision of DSCA is codified in DoD Directive 3025.18, which defines DSCA as, *"Support provided by U.S. federal military forces, National Guard, DoD civilians, DoD contract personnel, and DoD component assets in response to requests for assistance from civil authorities for special events, domestic emergencies, designated law enforcement support, and other domestic activities."* With a majority of CONUS RPA assets capable of responding to emergency situations on a short-term basis, the ANG has assumed, and will continue to be charged with, a leading role in ensuring the informational needs of the nation are met in times of emergency or disaster.

Figure 11: MQ-9 Reaper Conducting Touch-and-Go Landings During a Training Mission

Current FAA regulations, procedures and standards specifically addressing RPA flight operations restrict how, when, and where remotely piloted flights may occur in the NAS. While the force structure continues to grow to meet the needs of the Air Force and the joint force, efforts to achieve increased access to the NAS have not progressed to meet that demand. Current RPA access is limited by FAA regulatory rules that govern the operation of RPA in the NAS, and similar regulatory issues exist for flights within international airspace.

Without FAA-mandated safety mitigations, such as chase aircraft, or Certificate of Authorization (COA), RPA flights are limited to existing special use airspace (Restricted or Warning Areas). Operating exclusively within special use airspace and continued use of chase aircraft are not viable solutions to meet growing DoD requirements. Without routine access to the NAS, the capabilities of RPA and SUAS are degraded and the overall effectiveness of the U.S. military force is reduced. The Air Force will continue to work with FAA to develop material solutions to Airborne Sense and Avoid (ABSAA) to meet FAA requirements. Similar efforts must be leveraged for use of international and foreign national airspace through continued engagement with partner nations to enable UAS operations within GCC AORs. International airspace integration may require Office of the Secretary of Defense (OSD) involvement to possibly amend international agreements and Status of Forces Agreements. Access to foreign airspace or allied nation airspace is among the largest limitations to RPA operations.

3.   STRATEGIC ENTERPRISE VISION

This section, Strategic Enterprise Vision, discusses a macro-level, forward-leaning approach for the future state of RPA, focusing on general considerations for the future operating environment, missions, and capabilities. The following section, Section 4, Enabling Concepts, is intended to expand on this vision and delves into greater detail on the operational impacts and the S&T or R&D initiatives that are ongoing or required to support the envisioned capability growth.

The Air Force strategic vision is to meet future challenges through fully integrated airpower that includes critical RPA capabilities. NextGen RPA must be capable of performing a broad range of missions in the context of approved DoD scenarios, consistent with Air Force doctrine and the applicable joint publications. To this end, the Air Force must continue to invest in technologies that advance system performance and cross-domain capability. Such investments will result in procurement of a set of NextGen systems, including the NextGen Group 4 and 5 multirole aircraft, collaboratively networked with manned platforms to achieve ISR and kinetic and non-kinetic effects in the battle space.

> *"We must remain as committed as we were in 1945 to pursuing the most promising technological opportunities for our times, to having the scientific and engineering savvy to bring them to reality, and to having the wisdom to transition them into the next generation of capabilities that will allow us to maintain our edge."*
>
> *– Werner J.A. Dahm, Chief Scientist of the U.S. Air Force (October 2008–September 2010)*
> *Technology Horizons: A Vision for Air Force Science and Technology*

All new aircraft procurement, beyond those specified in the current Air Force programs, will consider both manned and remotely piloted configurations in the AoA or capabilities-based assessments (CBA). Beyond acquisition and life-cycle costs, the analysis must consider the associated benefits and limitations of each option, including combinations of manned and remotely piloted teams or optionally piloted capability. Reliability and sustainability of a desired capability are major considerations in the decision of whether manned or unmanned is the appropriate alternative. As an example, some questions to consider when evaluating manned versus unmanned options: Does the mission require extreme endurance? Does the mission require a high level of risk (whether environmental risk, such as a contaminated area, or operational risk, such as an integrated air defense system [IADS])? Does the risk and cost of losing an aircraft outweigh the mission it was designed to perform? These examples illustrate the importance of considering manned or unmanned alternatives where appropriate and pursuing the appropriate course based on analysis and affordability. This will allow the Air Force to achieve better strategic focus while maximizing cost efficiencies in a fiscally constrained environment.

Unmanned systems raise new issues of artificial intelligence (AI), communications, autonomy, interoperability, propulsion and power, and manned/unmanned (MUM) teaming that will challenge current test and evaluation capabilities and methods. These problems will get more serious as systems become more interactive and more automated. Failures often occur at the interfaces between systems elements, and in many cases, between interfaces originally designed to be separate. Current trends of interdependent software and network communications indicate that many elements of a system can now affect one another and that increased regression testing will be necessary. The incredible complexity inherent within millions of lines of software code requires new approaches for detecting problems earlier in the Design phase, where cost-mitigation efforts are most effective.

Today's RPA perform selected missions within permissive, and in some cases non-permissive, environments despite platform limitations. NextGen RPA must be multi-mission, adverse weather capable, net-centric, modular, use open architecture, and employ appropriate levels of autonomy. NextGen RPA should also be able to carry any standard payload within their performance envelope, with an adjustable explosive yield, and provide multimode seeker capability.

3.1 Future Operating Environment

The future operating environment will be one of constant and accelerating change. Economic, demographic, resource, climate, religion, geopolitical climate and other trends will engender local, regional, and global competitions. In this complex environment, state and non-state actors will find new and more deadly means of conducting operations in all domains; these may include weapons of mass destruction, suicide bombers, long-range and precise weapons and advanced cyberattack capabilities.

> *"Develop and deploy technologies, such as flexible and scalable encryption for reconfigurable sensors and fractioned platforms that will allow the operator to fight through an adversarial attack."*
>
> *– Air Force Cyber Vision 2025*

By no means is it certain that the United States and its allies will maintain their overall lead in technological development in the future. Enemies of the United States recognize the advantages of UA and will seek ways to mitigate and defeat such capabilities. Adversaries will continue developing formidable remotely piloted technologies that, in turn, require the United States to develop counter-UAS capabilities.

3.2 Future Missions

The operational environment will continue to evolve rapidly, requiring innovative operational employment concepts. Remotely piloted systems can help in countering emerging threats through the unique capabilities provided by these platforms. As capabilities improve and experience with RPA and SUAS increases, the missions for which each of these could be considered are expected to expand.

3.3 Future RPA Attributes

Future missions necessitate advances in many current capabilities and the introduction of new capabilities and technologies to meet future domain challenges. Advances will range from improvements in currently fielded systems to technology needed for future platforms. Attributes include sensor and C2 improvements, increased autonomy to reduce pilot workload and processes that allow for more efficient use of limited communication bandwidth. The ability to operate and maintain RPA and SUAS in adverse weather and improved MUM teaming will be paramount to defeating the growing threat. Teaming concepts should include the integration of a variety of unmanned systems, including ground, surface, subsurface and air systems.

3.3.1 Information Synchronization

As information needs and collaboration with other services improves, operations centers (e.g., SOC) must have the ability to ingest any data format; either generated from a tactical sensor on the ground or received from national assets in space/cyberspace, and use it for collaboration throughout the Air Force and the Defense Intelligence Information Enterprises (DI2E). Operations centers shall be capable of exchanging data with the intelligence community through the system high enclave and shall be capable of exchanging data at classification levels releasable to coalition personnel. In addition, operation centers should be capable of cross-domain transfer in accordance with applicable governing directives.

The RPA Operations Coordination Center (ROCC) must provide short-term storage capability for all GCS and platform sensor–derived data across all enclaves and allow operators shared access to data stored within the storage area network. Long-term cloud-based storage should be considered to permit sharing across operations centers and PED nodes. The ROCC (see Figure 12) must provide the capability for collaboration with other operation centers, C2 nodes, PED sites and other external agencies as required.

Figure 12: Notional RPA Coordination Center

The ROCC provides the main communications link between C2 and PED elements during operations. The capability must have secure voice and chat communications over applicable security enclaves with supported mission-relevant agencies as required for mission execution (e.g., intercom/intracom, telephone, chat).

The ROCC must provide its platforms with regular, frequent updates for mission execution. These updates must include weather, threat, target, track and mission status and capability information. Information is critical to platform safety of flight to ensure proper target identification and mission execution.

3.3.2 Exploitation Needs

Given the current manpower requirements of the PED process for data exploitation and the impact to both DCGS and squadron-level intelligence analysis, increased research and effort should be placed on developing tools to automate PED that would filter/screen massive amounts of collected video and audio and assist analysts in producing products. Tools should alert analysts to areas that require a more detailed or sophisticated analysis/processing effort based on set parameters. PED sites must provide the capability for collaboration with other PED sites, operations centers, C2 nodes and other external agencies as required.

3.3.3   Survivability

Current platforms have primarily operated in permissive environments, but in future conflicts, these assets must operate effectively in contested or denied environments. The Air Force must determine what capabilities are required and then conduct targeted cost benefit analysis to determine which are best suited as retrofit solutions for existing platforms vice a potential NextGen acquisition requirement. In broad context, NextGen RPA must detect, avoid, or counter known threats, via traditional and innovative means, to enable operations in a range of threat environments, from permissive to A2/AD, while maintaining a persistent presence over the target area. NextGen RPA and SUAS must leverage joint air and ground assets when ground-to-air threats exist. This may be achieved through a combination of high subsonic/supersonic speed, low-observable/low-acoustic technologies, operating altitude, maneuverability, employment of air-launched SUAS (AL-SUAS), active/passive countermeasures, or expendable assets.

3.3.4   Communications, Command and Control

RPA leverage the RSO architecture to enable global reachback and maximize capability while allowing for flexible mission execution. Each segment of the architecture (i.e., air, space, terrestrial and cyberspace) serves as a critical component in enabling agile, global capability. Current RPA communications architectures are highly effective but were designed and fielded rapidly with a platform-centric, closed architecture that was simply duplicated to meet urgent, growing CAP demand. Future demand on the communications infrastructure will require a secure, robust, agile, efficient, and redundant C2 and information distribution network. These improvements are essential for mission success in all environments and to ensure information delivery or guarantee positive C2. The efficiency will be enhanced through a common control system, control architecture and a dedicated C2 spectrum.

3.3.4.1   *Agile, Secure, Efficient and Robust Communications*

RPA communications planners must begin to consider numerous factors for fielding capability for future environments, including eliminating single points of failure and ensuring redundancies across all segments of the proposed architecture, availability of bandwidth (both LOS and BLOS), spectrum supportability, joint and coalition interoperability and networking, the need for low probability-of-intercept/low probability-of-detection (LPI/LPD) links in A2/AD environments and security needs across all environments.

The future for RPA must include an enterprise communications architecture, managed end to end with rigorous configuration control, sufficiently agile and robust to handle surges in demand globally. The architecture should be fully redundant (where practical) across all communications segments. The architecture must include a predominately Internet protocol (IP)-based aerial layer, aligned with the DoD Joint Aerial Layer Network (JALN), to move the full spectrum of data across the AOR in support of warfighter needs, including operations where satellite communications (SATCOM) is degraded. RPA must also be self-healing, wherein interruptions to a given segment are instantly absorbed by other redundant paths.

In the future, there must be secure, high-bandwidth communications capabilities that leverage modems readily adaptable to challenging and dynamic environments, potentially outside of the current traditional RPA radio frequency (RF) bands and leveraging dynamic spectrum reallocation. Spectrally agile communications capability must ensure spectrum flexibility and access to available spectrum bands worldwide to provide cyber resiliency and disruptive tolerant network access in the increasingly congested spectrum and contested electronic warfare (EW) environments. In addition, RPA should take advantage of emerging spectrum bands that will have internationally approved support for RPA C2 operations, including improved support for operations in the NAS.

> *"Emerging development of higher bandwidth components and devices has enabled the use of previously unused spectrum for communications, such as W band, 75–110 GHz, millimeter wave communications. These technologies further enable the development of simultaneous multi-mission, multimode spectrum sharing techniques."*
>
> **—U.S. Air Force Global Science and Technology Vision, 21 June 2013**

Current and near-term RPA data link protection capabilities through use of encryption require significant manual effort to implement and, once implemented, lack operational agility or transparency. In the future, enterprise-wide encryption capability must be simplified to make use of Public Key Infrastructure architecture to distribute key material to desired operating locations. In addition, distribution methods must be able to deliver key material and validate users through over-the-air keying to minimize the burden to the warfighter while protecting the network from intrusion or interception.

The following offer the potential to reduce costs or improve efficiency:

- Implement capabilities, such as bandwidth-efficient (BE) modems and data links
- Increase use of onboard storage/processing with appropriate consideration to consequence management plans in the event of a downed RPA
- Appropriately consolidate RPA communications infrastructure into a unified, centrally managed architecture for RPA global communications
- Use inclined orbit commercial satellites
- Purchase longer-term bulk commercial satellite leases to reduce total lease costs
- Increase the diversity of ground data site locations to allow efficient global coverage.

3.3.4.2   *Common Control Systems and Control Architectures*

The Air Force has begun collaboration with DoD and the other services on a joint concept for common control systems and control architectures that could serve as a model for appropriate future development of common control systems and architectures between services. This joint concept is intended to provide direction to the services and summarizes near-term, mid-term and far-term efforts to move toward a joint solution for RPA control systems and architectures.

> *"Acquisition of proprietary systems is costly and can impede interoperability and the reuse of components among systems. Conversely, an open system that incorporates modular design and open standards for key interfaces can readily accept upgrades from a variety of suppliers without redesign of the entire system, providing numerous cost, schedule, and performance benefits. "*
>
> **—Open Systems for UAS, Government Accounting Office Report GAO-13-651**

To meet interoperability requirements and enable envisioned efficiencies, the services must work together to—

- Publish a joint, non-proprietary commercial off-the-shelf (COTS) framework, configuration descriptions and interface definitions to efficiently allow industry partners to develop common architecture services and applications.

- Establish joint governance procedures, policy and oversight of the COTS framework, and interface definitions applicable to RPA control systems and architectures.
- Jointly develop common internal/external interfaces and standards (e.g., Universal Systems Interoperability Profile [USIP], RPA Interoperability Profiles [IOP]) that will promote interoperability of control systems.
- Ensure development of control system software services and applications is aligned with the common architecture strategy and leverage existing efforts, including UAS Control Segment (UCS) and UAS C2 Initiative (UCI), where practical.
- Build control system software services and applications to a common architecture model with standardized interfaces and developed with government data rights.
- Develop and field control systems with maximum reuse of services, applications, and common components, thereby reducing redundant development efforts and total ownership costs.
- Apply common control standards and interfaces across the family of RPA.

3.3.4.3   *Dedicated C2 Spectrum*

One significant limitation to RPA operations in the CONUS and outside the continental United Sates (OCONUS) is the lack of spectrum supportability for C2 data links. The use of C-Band has been a pervasive problem that impacts mission accomplishment because C-band for RPA must operate on a non-interference basis. Other similar scenarios have the potential to halt operations until the spectrum issue can be resolved. In the future, the vision includes establishing a spectrum band dedicated to RPA C2, where RPA have primary status and preferably operate in a band that works closely with existing RF equipment requiring minimal configuration changes and corresponding cost. This will require close coordination between Air Force requirements, SPOs, and Air Force, DoD, and other U.S. government spectrum representatives of the U.S. delegation to the International Telecommunications Union and the World Radiocommunication Conference (WRC).

3.3.5   Interoperability

Airpower requires nearly instantaneous coordination of action, clear communication with the commander, and support that provides what is needed at the right time. The Air Force achieves airpower through a networked set of systems operating across the seams of air, space, and cyber domains and not just as a collection of individual platforms. All too often, synergistic capability is not achieved until after the platform has been fielded because its capabilities were developed and procured in isolation. It is this creation in isolation, or stove-piping, that has resulted in redundant costs and delays in integration of those capabilities for the warfighter. The key to unleashing the full potential of the RPA is interoperability.

The joint UAS theater CONOPS (see Figure 13) supports the notion of operational flexibility provided through interoperability to support requirements across the spectrum of conflict, from centralized theater priorities to responsive direct support of ground maneuver units. This approach to interoperability considers organizational structure, C2, PED, weapons employment and airspace control. The CONOPS should maximize capability available to the joint force commander (JFC) or supported unit, considering all available assets. In an A2/AD environment, this CONOPS must enable effective air defense against adversaries.

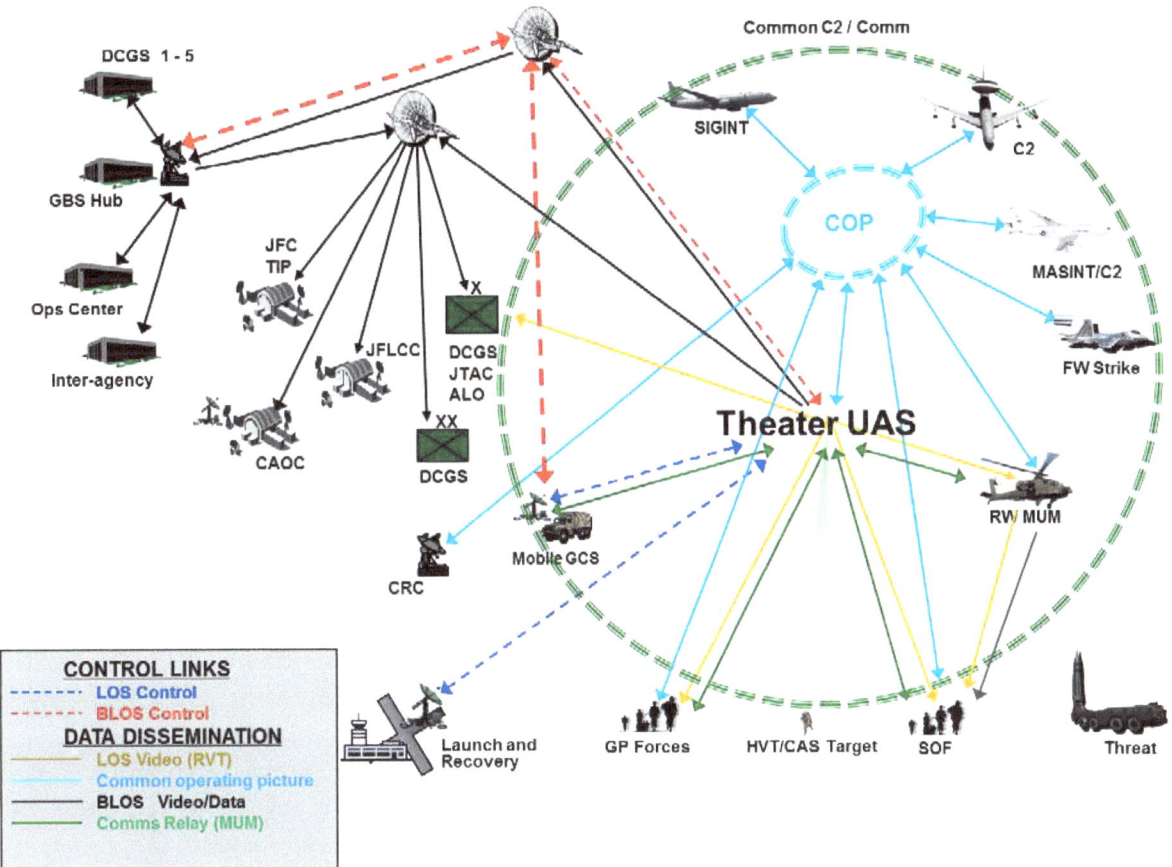

Figure 13: Joint UAS Theater CONOPS

The Air Force vision for interoperability is to achieve horizontal integration and commonality across remotely piloted, manned and ground support systems. It is embodied in the FoS development approach (see Section 3.4). The ultimate goal is to improve capability for the warfighter while simultaneously reducing life-cycle costs, enabling future concepts, and increasing system interoperability across the joint operational environment. To accomplish this vision, specific standards and interoperability profiles for data, data links, and service-oriented architectures (SOA) must be developed and implemented across the enterprise. The Net-Ready Key Performance Parameter has recently been updated to include a requirement to test as a FoS rather than simple compliance with a broad set of standards. This may address some interoperability issues for large programs, but smaller programs with less oversight will need to implement similar controls.

As a first step, the Air Force should conduct a detailed CBA to recommend prioritized investment to meet the requirements set forth by the CFLIs. This should also consider guidance set forth in the Air Sea Battle Concept and to meet the demands of a potential A2/AD conflict. This effort is intended to inform other investment strategies, and the Air Force should consider interoperability in future requirements, though it is understood that this must be balanced with urgent warfighter requirements.

Some of the other actions the Air Force is pursuing include developing specific requirements, initiating oversight, mandating a common C2 SOA for GCS and incorporating modular payload interfaces.

3.3.6 Sensors

NextGen RPA must employ sensors with increased range and sensitivity in a wider variety of environments. Improvements in sensor capability will improve mission effectiveness and aircraft survivability, allowing for greater standoff ranges and reducing impact of obscurants and adverse weather. Further, sensors must be developed to provide airspace SA and may contribute to increased safety for airspace integration into civil and congested combat airspace. Improved sensor reliability (e.g., mean time between failures) could enable growth in high-altitude, long-endurance (HALE) platform capability. Similarly, miniaturization could ease the size, weight, and power (SWaP) limitations for SUAS and could enable RPA to carry multiple or more sophisticated sensor payloads. Designers will have the added challenge to provide plug-and-play compatibility/drop in capability between RPA models while reducing the SWaP of the sensor payloads. Lastly, the ability to cue a particular sensor quickly and simply using data provided by a different onboard sensor or offboard data source will be a vital capability for the NextGen RPA. To support the expendable version of the AL-SUAS, disposable sensor technology needs to be developed.

3.3.7   Multi-Mission / Modular

The lessons learned from current RPA missions and employment, such as the MQ-1/9, in support of recent combat operations highlight the necessity for multi-mission capabilities. The Air Force has fielded several quick-reaction capabilities (QRC) to meet urgent combat needs, resulting in more than 20 uniquely configured MQ-9 aircraft. This is logistically unsupportable in the long term. To address this, the SPO is integrating modular interfaces. The MQ-9 will implement the Universal Armament Interface (UAI) developed by the Air Armament Center for manned fighters and bombers to allow integration of new weapons without rewriting the aircraft operational flight program. This same concept is now being used to develop a Universal Sensor Interface to reduce the number of unique aircraft configurations and facilitate future QRC without aircraft modification. In the future, Groups 3–5 UA should be capable of supporting multiple missions with multiple payloads simultaneously (e.g., weapons, communications relay pod, radar). Platforms must be tailored with capabilities shaped to the mission needs of the supported commander and allocated as needed while airborne throughout the AOR.

Modularity is the ability to mix and match weapons and sensors to meet given mission requirements on a platform. Incorporating the UAI will be a potential solution for weapons modularity. Modularity is the key enabler for RPA mission agility, flexibility, and adaptability and growth capability to support expanded missions. Modularity can lower costs by providing a way to upgrade, augment or replace technologies while preserving the bulk of the initial investment. Efforts are underway to develop modular payload interfaces that will be implemented by both the MQ-9 and possibly RQ-4 to reduce the number of platform configurations and save integration costs. Beyond the limits of current DoD research, development, testing, and evaluation (RDT&E), developing a modular system is a way to leverage other discoveries and developments.

3.3.8   Operational Reconnaissance

RPA equipped with gateways are ideally suited to support Operational Reconnaissance operations. Currently they are being equipped with a link gateway system, including the Vortex-based BLOS-C2 on MQ-9s and BACN on RQ-4. These systems can connect to Link-16 and the other tactical data link networks. Operational reconnaissance collections performed by platforms with tactical data link connectivity to these pods can then link into the high-capacity backbone data link architecture and potentially leverage RPA RSO. As this concept matures, the close working relationship between RPA systems and intelligence organizations provides established RPA aircrew reachback for Operational Reconnaissance. An aircrew on a strike mission could provide sensor data gathered from the primary mission to further characterize the battle space, while ground-based operators simultaneously have access to the entire common operational picture (COP).

3.3.9 Adverse Weather

NextGen RPA must be able to operate in adverse weather and environmental conditions. The first step in this process is to evolve current RPA to be at least as capable as today's manned platforms. Future airframes must be weather tolerant, operate above and below the weather, and be able to rapidly climb or descend through the weather. NextGen RPA must be able to execute missions (both sense and engage) in extreme weather conditions and adverse environments (e.g., day or night conditions, jamming, areas of dense foliage/vegetation, enemy obscuration) as well as in all civil and military use airspace. NextGen SUAS will be required to maneuver and perform missions BLOS in urban environments and inside buildings, canyons and caves.

3.3.10 All Environment

NextGen RPA must be networked and integrate with a C2 system that collects information from a variety of sources, prioritizes information flows, and disseminates information to and from a variety of customers. This includes a robust, multi-level security environment that generates, collects, prioritizes, and assigns tasks to the appropriate C2 elements. NextGen RPA must also develop and disseminate SA feeds for all internal elements and external customers of NextGen RPA products. NextGen RPA must be capable of using both LOS and BLOS secure communications for aircraft control and able to collect and disseminate information that is completely interoperable and integrated with joint forces.

3.3.11 Weapons

NextGen RPA weapons must share many of the same characteristics as their sensors. The weapons should provide increased range and accuracy in a wider variety of environments. NextGen RPA weapons must also offer the flexibility for multiple employment tactics using onboard/external sensors, laser designator and GPS. In most cases, the NextGen RPA will employ many of the same weapons as those of manned aircraft. The platform must provide the flexibility to tailor weapons payload and employment to the mission.

3.3.12 High-Altitude Long-Endurance

The increasing demand for ISR and other airborne asset support and the high cost of space-borne systems has driven both the commercial and military sectors to examine HALE aircraft as a tailored solution between air and space operations for a range of challenges, especially in the ISR and communications domains. The capability should be balanced against service and GCC priorities to examine if the needs can be met through improvements to existing platforms. Long endurance is enabled by low and modest flying speeds. Station keeping in winds is a key design driver for very long and extreme endurance. Very long endurance is enabled by high-efficiency combustion engines (diesels or hydrogen) and fuel cells, very large aerodynamic spans and laminar flow wings. Extreme endurance also requires high-efficiency photovoltaics and fuel cells, high energy-density batteries or regenerative fuel cells, and very low wing loadings. HALE aircraft, by definition, fly above today's controlled airspace where most weather and current threats have less impact, and some potential platforms show promise for persistence ranging from 5 days to as long as a year without refueling. In the next 25 years, with considerable advances in technology, it may be possible that RPA could be capable of long endurance at high speed.

Significant technological hurdles remain for HALE aircraft. Expendable energy sources, such as jet propellant 8 (JP-8)/Jet A or liquid hydrogen, may be capable of endurance approaching a week. Regenerative and energy-harvesting energy technologies, such as solar thermal harvesting or photovoltaics, show potential persistence up to several weeks or even multiple years of endurance. Given the operating altitudes and commensurate propulsion choices, HALE aircraft require atypically low empty-weight fractions, which create significant challenges for lightweight materials and structures, aerodynamics, and aero-servoelastic design. Due to power and thermal design challenges (e.g., hybrids, fuel cells, electric), reliability is another enormous challenge for these platforms. In addition, potential weight and power considerations for high-altitude sensors must account for the need to harden components/materials for near-space environments. These are further compounded by the need to operate continuously for days, weeks or even years without maintenance.

If these technologies and system challenges can be successfully overcome, it is not difficult to imagine a day when specially equipped HALE aircraft can replace cellular towers over a natural disaster area or become part of the GPS constellation to help mitigate threat situations or constellation failures. In the future, the HALE capability could provide extreme persistent ISR capability over large areas, such as the Horn of Africa, to monitor pirate threats to commercial sea lines of communication (LOC), or serve as a near-perpetual battlefield communications node similar to the BACN capability on Global Hawk. Currently, the HALE Wingman and HALE Bomber concepts show potential for HALE platforms to carry weapons.

3.3.13 Efficient Engines/Alternate Power

Since the initial introduction of RPA and SUAS into the U.S. armed forces more than a decade ago, technological advances in propulsion and power may offer new alternatives to traditional power solutions for platforms. Propulsion and power are key enablers to offer increased capability for RPA and SUAS platforms of all sizes and purpose.

For larger systems, excluding HALE platforms, current payloads and payload architectures are beginning to push the limits of power and thermal capacity. As architectures emerge that allow rapid payload integration and changes from mission to mission, power and thermal problems will only get worse as warfighters take advantage of the payload flexibility. Low-observable platforms will present a more difficult challenge due to extreme packaging constraints and reduced ability to dissipate heat into the environment. Adding thermal (e.g., phase change wax) and power (e.g., ultra capacitors, fuel cells) capacitance between the payload and the vehicle's power and thermal systems will help create a flexible power and thermal interface to support flexible payload requirements. Future platforms must consider creative heat-rejection techniques, such as rejecting heat into the engine bypass stream or other apertures like the aircraft's entire external skin. Variable engine cycles will also be a hallmark across all platforms, manned or remotely piloted, to allow for simultaneous performance and efficiency improvements across the entire flight envelope.

For small systems and HALE platforms, propulsion and power concepts begin to take on new dimensions. The inexorable march of miniaturization across multiple domains has led to increasing capability for a given size or the same capability in ever decreasing sizes. Miniaturization has led to individual units or allowing personnel to carry their own air assets in the form of SUAS. Payload, propulsion, and power scaling are complex endeavors, but forward progress is being made in these areas:

- Efficiency and noise challenges with scaling force new internal combustion designs, such as the nutating and rotary engines
- Efficiency and noise challenges of propellers at smaller sizes
- All-around reliability and weight, as components get smaller and the increasing importance of driving empty-weight fractions down at ever decreasing sizes
- The desire to use Jet A fuel with additives for most propulsion requires new compact fuel cell designs, finding solutions for high peak torque, and miniaturized injectors and fuel pumps

- The square-cube law forces designers to reject volume as a solution for component packaging, energy and thermal capacity, and power distribution and instead rely on structurally integrating those same features into the internal and external surface area of the vehicle or finding orders-of-magnitude improvements in energy and power density especially at the micro and nano scales
- Hybridizing to achieve all-around efficiency across a wide range of operating conditions.

The end result in the foreseeable future will be RPA that are able to accommodate ever-increasing payload flexibility requirements and demanding payloads, such as AESA radars and directed energy weapons in ever more challenging environments. In addition, SUAS and micro air vehicles will increasingly begin to provide capabilities similar to RPA.

3.3.14 Optionally Piloted Aircraft

An optionally piloted aircraft (OPA) is either designed from the beginning or is a retrofitted manned platform allowing all its systems to be controlled from human inputs via an onboard cockpit or remotely piloted. An OPA differs from a traditional RPA in that it must include onboard provisions for a crew—a cockpit, associated life support, and accommodations of key human factors and survivability concerns. Likewise, an OPA differs from a traditional manned aircraft in that it must have the necessary systems to allow for remote or semi-autonomous operation. OPA design should consider minimizing impacts to range, detection, and cost of acquisition.

From an operational standpoint, OPA are compelling where the remotely piloted mode expands the commander's options by operating at the extremes of range and endurance or without endangering aircrew. Commercial air cargo carriers are investing in research into OPA to support reduced crew operations and formations of aircraft on transoceanic flights. OPA also offer a possibility where the platform can perform some but not all mission requirements in a remotely piloted mode. OPA offer necessary flexibility where the use of manned or remotely piloted modes is not limited by a single point of failure, due to either enemy action or policy restrictions, and where it gives commanders freedom to apply force at various risk thresholds. In addition, the potential exists to allow modest reductions in pilot manning and training/readiness requirements by leveraging automation and autonomy that permit multi-aircraft operations or efforts to reduce the one-to-one aircrew-to-aircraft ratio for select missions.

3.3.15 Autonomy

Air Force missions increasingly involve tasks that must be accomplished on a scale beyond human capability. As such, it is reasonable to reconsider how we look at autonomy for combat missions. Some combat decision cycles occur at speeds that are many orders of magnitude faster than human reaction time. Systems will need to automatically respond, nearly instantaneously or at a very precise time, to achieve a desired effect. More appropriately, autonomy should be applied in difficult, dangerous, and monotonous scenarios. This concept can be illustrated by flight control technology on modern dynamically unstable aircraft. Pilots of modern fighters cannot observe, orient, decide and act (OODA) quickly enough to control the aircraft, so the computer executes the actions necessary to achieve the desired effect. Collaboratively, the pilot and the flight control computer fly the aircraft to perform the mission. By extension, properly applied autonomy will increase the human SA while performing computation at speeds beyond human capability. Likewise, systems today collect volumes of data that exceed the Air Force's capacity to review in a timely manner. Identification of relevant cues in this vast amount of data requires vigilance well beyond human abilities. Autonomous correlation systems could search data collected and nominate potential targets. Employing unblinking autonomy to sift through the data will enable personnel to concentrate on translating processed data into information and making decisions based on that information. In these examples, autonomy does not replace humans but rather changes the way humans do tasks while exponentially increasing their effectiveness. People overseeing autonomous processes and systems collaboratively teamed together have the potential to revolutionize warfare, particularly when applied to RPA.

The Air Force vision for autonomy is to increase warfighter effectiveness by enhancing remotely piloted systems capabilities and expanding their capacity to create effects in the battlespace. The initial focus must be on onboard processing of data from sensors to reduce bandwidth required, further enabling human analysts to be more efficient and effective by focusing their efforts on key information. Autonomy with well-designed human-systems integration (HSI) must also integrate flight and mission information to enable true multi-aircraft control for transit operations. Control of multiple RPA in the near term by a single crew requires a high level of autonomy, no lethal effects and a permissive environment. In addition, the targets must be immobile, and the essential elements of information cannot require surveillance or reconnaissance of activities or personnel associated with a target. With appropriate CONOPS and doctrinal considerations, the future potential for autonomous systems to independently select and attack targets with lethal effects exists from a technology perspective. To achieve this, the Air Force and DoD must first address the legal, moral, and ethical concerns that autonomous lethal effects present as well as consider minimum safeguards. Future RPA may continue on mission using a combination of autonomous behaviors and assured position, navigation, and timing (PNT) if communications are lost or degraded in an A2/AD scenario.

As this advances over the next 15–20 years, autonomy will mitigate loss of data links and enable loyal wingman aircraft to fully integrate with manned and remotely piloted platforms through LPI tactical data links. The near-term concept of swarming consists of a group of semi-autonomous UAS operating in support of both manned and unmanned units in a battlefield while being monitored by a single operator. Loyal wingman technology differs from swarming in that a UAS will accompany and work with a manned aircraft in the AOR to conduct ISR, air interdiction, attacks against adversary IADS, offensive counter air missions, C2 of micro-UAS, and act as a weapons "mule," increasing the airborne weapons available to the shooter. The system should be capable of self-defense, and thus, a survivable platform in contested and A2/AD environments. The loyal wingman UAS could also be a "large" UAS that acts as a cargo train or refueling asset. Simultaneously, machine-to-machine interfaces and advanced collaborative SoS control software will enable swarms of SUAS to function in a self-forming rule-/role-based airborne network to create virtual large array antennas or cyber and kinetic effects that over-saturate an adversary's defensive systems at a relatively lower cost.

Figure 14: Notional UAS Teaming

Within 25 years, autonomy will accelerate the OODA loop to provide critical information to decision makers orders of magnitude faster than humans, this will be crucial in future combat scenarios against a high-tech adversary. The RPA Vision advocates for continued investment in autonomy, because technology advances are required to maintain an affordable force structure and effectively achieve airpower in an increasingly complex environment.

3.3.16 Airspace Integration

The Air Force requires routine RPA access to the appropriate airspace required to meet mission needs. A robust integration CONOPS for all classes of U.S. airspace will enable future RPA to operate with manned aircraft, will provide seamless interaction with air traffic authorities and airspace regulators, and is fundamental to flexible worldwide RPA deployment.

> *"...it is vital for the Department of Defense and the Federal Aviation Administration to collaborate closely to achieve progress in gaining access for unmanned aerial systems to the National Airspace System to support military requirements."*
>
> *– 110th Congress, National Defense Authorization Act for FY09, Sect 1039*

The DoD is working near- and mid-term solutions allowing for immediate gains in NAS access, while working toward viable long-term solutions. Priority is given to initiatives that reduce the constraints required by the FAA and expedite the approval process. Near term, the DoD is expanding current military airworthiness guidance to address those component and system attributes that are unique to UA, establish training standards, and address other regulatory alignment. Mid-term, the Air Force intends to develop and certify ground-based sense and avoid (GBSAA) solutions to meet a sense and avoid (SAA) requirement to gain localized access. The long-term vision of full integration in all airspace, whether under civil or military control, will be to implement technologies, such as ABSAA and NextGen ATC system technologies, such as Automatic Dependent Surveillance-Broadcast (ADS-B). This will allow seamless integration without a need for special RPA or SUAS airspace control measures or segregated airspace. A fully certified ABSAA solution has the potential to revolutionize aviation safety, as it can be immediately available for application on both manned and remotely piloted platforms.

3.4 Family of Systems

The asymmetric game-changing attributes of RPA and SUAS impact all levels of conflict. To balance future investments, the Air Force has a FoS approach to developing and integrating capabilities to collectively support the missions. AFSOC is the Lead Command for Groups 1–3 UAS, while Air Combat Command (ACC) is the lead command for Groups 4 and 5 UAS (see Figure 15). There are also specific sets of missions that will have unique platforms or interfaces, such as stealth, hypersonic or long endurance.

Currently, SWaP capacities align most mission capabilities. The larger the aircraft and more robust user interface, the wider the set of mission capabilities an individual UA can provide. Today, most RPA and SUAS are flown individually with limited integration of other assets. Collectively, the FoS will evolve from cooperation to collaboration with other airpower assets and eventually incorporate capabilities to achieve cross-domain effects. To do this, the Air Force must synchronize FoS milestones beginning with interoperability and mission data sharing and leading to machine-to-machine mission interfaces between manned and remotely piloted aircraft in large force packages. This is critical as the FoS shifts from IW optimization to A2/AD scenarios.

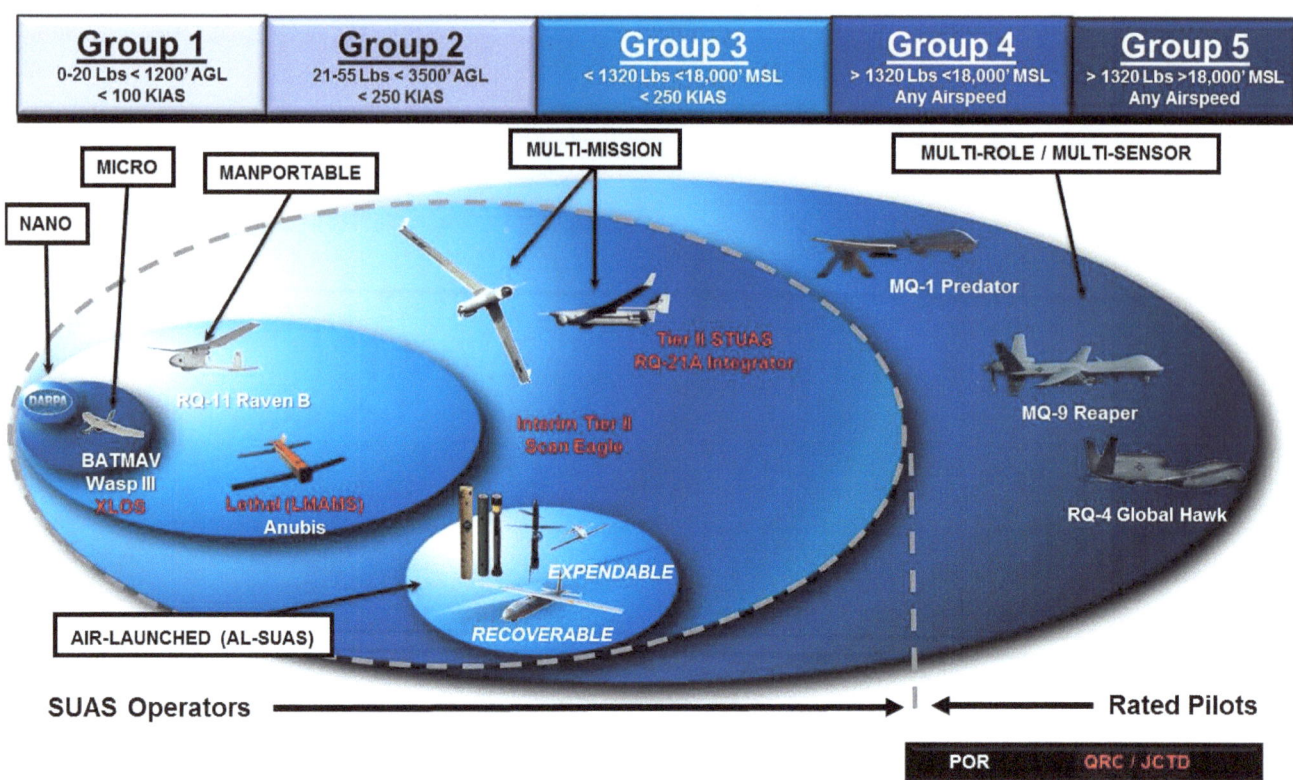

Figure 15: UAS Family of Systems

The asymmetric, game-changing capability of these systems impacts all levels of conflict. Specific aircraft, payloads, and control interfaces impact mission capabilities. All groups of UAS must meet specific interoperability requirements to support the joint warfighter. Synchronization efforts contain key steps and milestones affecting the entire Air Force RPA spectrum of capabilities. SUAS must be integrated to support IW while continuing preparation for a near-peer A2/AD threat. SUAS and RPA must continue to integrate with manned aircraft operations through teaming concepts to prepare for future major combat operations and contested environments.

The Air Force must continue the FoS approach to developing and integrating capabilities to collectively support the missions. In general, the smaller the vehicle, the simpler the user interface and the more specific the vehicle mission capabilities. Larger platforms require a more robust user interface and offer options to support integration for a wider set of mission capabilities. There are three FoS that cover the capability vision for the majority of systems flown. Currently, relatively small quantities of niche capability systems are being procured.

3.4.1   Group 1–3 UAS FoS—SUAS

The Air Force must leverage the AFSOC vision and lessons learned to develop a long-range plan for acquisition of Group 1–3 FoS—SUAS and pursue full integration into the service's MUM teaming concepts. The Air Force's SUAS capability can integrate seamlessly into a variety of mission areas. However, without a more concentrated and deliberate focus, the service will not fully realize the capabilities that SUAS could bring to support overall Air Force mission sets.

Due to their expendability and lower costs of acquisition and training, SUAS will play a key role in warfare, including emerging counter-UAS missions. Large numbers of lethal or network attack SUAS may be used to saturate enemy defensive forces and potentially cause them to expend a large portion of their kinetic weapons to reduce the threat to U.S. and coalition forces in an A2/AD scenario. For IW, it is possible that the next inexpensive asymmetric threat will be a SUAS (e.g., an airborne improvised explosive device [IED]).

Examples of warfighter needs that SUAS could contribute to include—

- Capability to engage fleeting targets with low collateral damage
- Capability to detect, precisely locate, identify, track, transmit coordinates and assess attack results against fleeting targets below weather
- JTAC need for BLOS targeting
- Expeditionary/quick reaction force ISR.

The SUAS community faces many of the same challenges as RPA systems: a fiscally constrained environment and reduced OCO funding. Within the Air Force, there is a need for dedicated advocacy across the Headquarters Air Force (HAF) staff working in conjunction with Headquarters AFSOC, as the CFLI for SUAS capability development and integration. In addition, there is a need for continued development and advocacy of SAA systems for use by SUAS to expand access to training airspace and ensure operational use within CONUS is not restricted.

> *"Advances in multifunctional nano-electronics and nano-materials for low-cost and sustainable energy can provide another 100X improvement in size, computing performance, and power efficiency over the next 15 to 25 years."*
>
> **– U.S. Air Force Global Science and Technology Vision, 21 June 2013**

Group 1 UAS: Group 1 UAS are capable of conducting a variety of indoor and outdoor reconnaissance sensing missions using micro-electronic machines technology. The system is intended for operations in confined spaces, directly supporting a small unit and not intended to be employed for persistent ISR at altitudes above 1,200 feet above ground level. The UA should be less than 450 g in mass, the maximum airspeed and frangibility should minimize damage by unintended collision with personnel or vehicles. Advanced development of the Group 1 UAS will introduce capabilities never before realized. These include the ability to perform surveillance missions inside buildings and in confined spaces.

In A2/AD environments, air-launched nano/micro SUAS may be the best means to provide persistence at a specific location. Technologies need to be developed to allow nano vehicles to "perch," collect, analyze, and communicate at very low power levels. Perching missions may include collections from nano-cameras, acoustic, and SIGINT in the near term. New battery technology, solar harvesting, and the ability to "borrow" from the adversary's power grid need to be researched and applied to this group of vehicles. Further, the use of bio-mechanical technologies will require legal and doctrinal development on how these potentially lethal systems are employed.

Group 2 UAS: Group 2 UAS address the need of small battlefield airmen teams for a more robust, greater endurance, mobile, man-portable system carried by the individual team in either mounted or dismounted operations. These systems can be used to sense, engage and destroy threat targets with focused lethality at close ranges within 10 km. This group is a key capability used by Air Force security forces teams to secure air bases at domestic missile/space launch sites and forward locations.

Group 3 UAS: Missions requiring increased SWaP over Group 2 UAS but reduced operating costs, reduced forward-deployed footprint, and faster speed of deployment and initial operations (compared to RPA) can be fulfilled by Group 3 UAS. All of the Group 2 benefits of multiple disparate payloads, multiple networked UAs (and manned aircraft), plus the ability to carry and employ weapons under cloud decks and longer individual aircraft endurance accrue. This particular group of UAS is of value in building partner capacity to observe and control the ungoverned spaces of countries that the United States has an interest in supporting.

The asymmetric game-changing capability of Group 3 UAS impacts all levels of conflict. In scenarios where the military has forward-launch capability, Group 3 systems can play a key role in directly supporting SOF teams, other ground units, and manned assets by providing armed overwatch and communication information, engaging more targets, providing decoys, and jamming and disrupting enemy attacks. Multi-mission SUAS also support the Air Force as a tool for the building partner capacity mission.

Air-Launched SUAS: AL-SUAS are aircraft that address the need for offboard sensing from manned aircraft and RPA. These can be controlled from the parent aircraft or surface teams trained to operate them. AL-SUAS provide the flexibility to conduct offboard sensing missions, focused lethal engagements, and multiple diverging target tracking. Air-launched capability includes both expendable and recoverable assets, which provide unblinking eye coverage. Joint doctrinal shifts may be needed to address how AL-SUAS are employed. Past lessons should be applied to use of AL-SUAS to enable more effective MUM defensive counter air, suppression of enemy air defense (SEAD), and special operations missions. This will require development of appropriate autonomy to enable manned aircraft to direct AL-SUAS missions via tactical data links and maintain precise relative positioning for electronic attack (EA) and SEAD. Swarm technology will be used to deconflict flight paths while collaborating on mission execution for multiple AL-SUAS. Swarm technology will allow the commander to virtually monitor the UAS both individually and as a group. A wireless ad hoc network will connect the UAS to each other and to the swarm commander. The UAS within the swarm will fly semi-autonomously on a pre-programmed route to an area of interest (e.g., coordinates, targets) while also avoiding collisions with other UAS in the swarm. These UAS will automatically process imagery requests from low-level users and will "detect" threats and targets through the use of AI, sensory information, and image processing. Swarming will enable the UAS network to deconflict and assign the best UAS to each request.

By acting as offboard sensors, AL-SUAS can increase the ISR capability of manned and remotely piloted aircraft. These offboard sensors can provide ISR at stand-off ranges, go below the weather, and follow multiple diverging targets. In addition, SUAS inherent LPI/LPD characteristics could be built upon for a future denied area penetration capability. These can be controlled from the parent aircraft or ground teams trained to operate them.

AL-SUAS can also provide unique lethal strike capability to manned and remotely piloted platforms with precision, low collateral damage, and a live video feed of the final link in the kill chain. This capability was recently proven by the ground-launched Anubis prototype. In addition, AL-SUAS provide flexibility that is not found in current precision weapons. The ability to loiter, engage, wave off, loiter and re-engage is unique to this type of SUAS.

Finally, AL-SUAS could provide increased capability to ground teams. When working with a "stack" of aircraft, AL-SUAS can be launched and handed off to those teams for organic ISR or strike missions. This concept could reduce the weight that a ground user must carry. The expendable, or optionally recoverable, nature of AL-SUAS does not add unnecessary complexity to missions.

However, for an AL-SUAS CONOPS to work, the SUAS must be designed to reduce the impact on a heavily tasked operator (see Figure 16). Multiple SUAS in the airspace supervised by a single operator or multiple operators is technically and procedurally challenging. This further complicates airspace control and air battle management for those who are responsible for coordinating and integrating dynamic maneuvers and attacks. The challenges of controlling multiple SUAS simultaneously are currently being met by several cooperating development programs, but further study of C2 systems, processes, and organizations is required to ensure these issues reach positive resolution. Adaptable levels of autonomous operations offer a potential solution to these challenges.

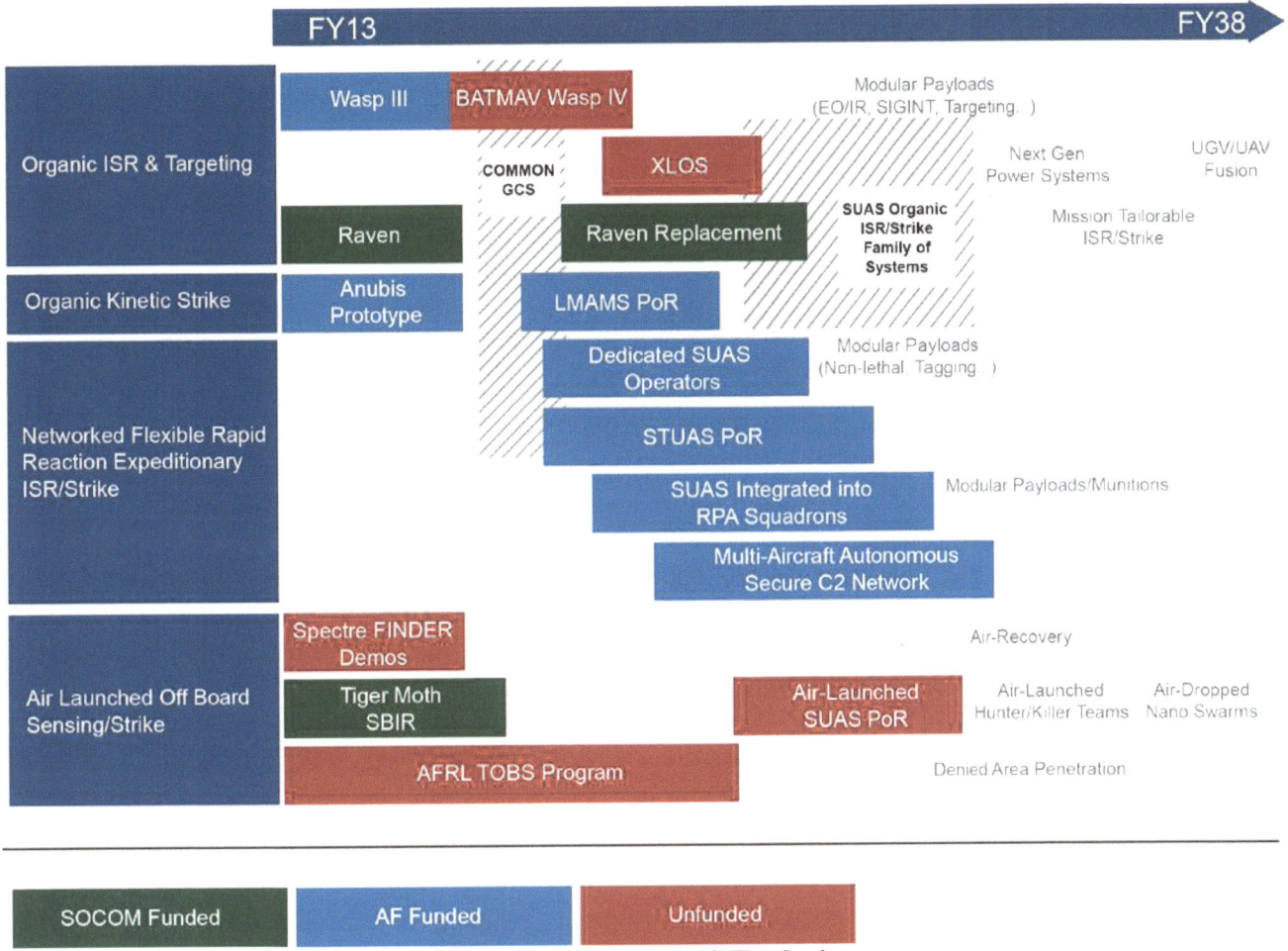

Figure 16: Small UAS Evolution

3.4.2   Group 4–5 UAS FoS—RPA

Group 4 UAS: Current Group 4 RPA systems have well-defined near-term capability improvements validated by the Joint Requirements Oversight Council (JROC). New payloads (podded and internal), encryption, data link evolutions, and open architecture GCS are programmed within the future years defense program. Improvements and modifications to current platforms and sensors must consider the ability to operate in adverse weather and increased threat environments. Emerging missions for Group 4, such as CAS, ballistic missile defense, and C2 relay functions may require reprioritization of new capabilities. These include auto takeoff and land, multi-aircraft control (MAC), remotely accessible IP network (RAIN), UAI, AL-SUAS, ABSAA, combat aerial ISR and tactical data link integration. The systems will use imported data from wide area search platforms, fused to a COP feed, and will provide SA and augment initial target/survivor cueing. Future RPA operating outside the threat envelope will provide rapid "on-call" precision weapons to accurately employ supporting fires when needed, as well as coordinate other supporting fires in support of joint force air component commander objectives.

NextGen RPA may be employed stand alone or in concert with other assets as a part of teams of other remotely piloted or manned aircraft. These teams will require machine-to-machine interfaces, spectrally agile data links, advanced target designators and markers and secure voice communications. These multi-aircraft MUM mission packages will be directed by current and qualified pilots with access to available SA information. These aircraft will be capable of operation in all weather conditions. Future analysis will determine the affordable tradeoff of standoff defensive systems and stealth to achieve required effects in an A2/AD environment.

NextGen RPA may augment relief and reconstruction (R&R) missions, PR, and stability operations with the capability to provide constant survivor status and geo-location data through advanced sensors and the endurance and persistence of the RPA. NextGen RPA may support combat aerial delivery operations with the capability to provide ISR for the drop zone battlespace, as well as detecting and neutralizing potential threats to combat aerial delivery aircraft and drop zone recovery personnel. NextGen RPA may also support the R&R missions by detecting and identifying potential threats to the recovery assets and survivor, as well as the capability to neutralize threats through lethal and nonlethal means. Given the expected end of service for MQ-1, an increased focus on NextGen RPA technology is imperative to invest appropriate resources to meet program objective memorandum and acquisition life-cycle requirements to meet initial operating capability in the 2030 timeframe.

The Air Force vision for a Group 4 RPA FoS is a modular, open architecture and networked system built around a common core airframe and should demonstrate some autonomous behaviors. This aircraft can be tailored with capabilities shaped to the mission needs of the supported commander and allocated as needed throughout theaters. With RSO and open architecture C2, global employment of any of these aircraft from a family of C2 interfaces worldwide will maximize capability available to the joint force at the lowest personnel and equipment cost. The aircraft of the future should incorporate modular payloads such that new capabilities can be integrated without redesign of the platform. An open architecture interface for weapons allows air-to-ground and air-to-air weapons and AL-SUAS employment from current and future weapon/sensor inventories. This FoS may include an optionally manned capability. The global distribution of responsive and flexible multirole RPA will serve CCDRs with a networked scalable capability and a minimum forward footprint.

Each aircraft will be flown from an advanced family of distributed control stations. In the near term, the Air Force must invest in multi-aircraft-capable open C2 architecture GCSs with the ability to hand off some mission tasks to an RPA C2 node or mission router. As soon as a broader governance structure can be commissioned, the various service architectures will be merged into a common or universal architecture. The sooner this happens, the more savings will be realized through shared C2 services. As this matures, human interfaces will be optimized for the mission phase or type of mission with rapid, seamless handover of control between the families of control stations. Transit operation workstations may be used to control formations of UAS simultaneously, as determined by CONOPS and the ability of the pilot to manage the workload safely and effectively. Methods to control multiple UAS simultaneously must consider control across various groups (e.g., Group 4 UAS controlling Group 1–3) and levels of complexity. Missions will be flown from GCSs leveraging technologies derived from HSI and human-machine interface (HMI) research and technical solutions must be developed in close coordination with DoD and FAA governance and policy organizations.

In addition to airworthiness certification benefits, separate IP addresses for payloads and aircraft control may have operational benefits as well. Distribution of sensor control and data, selectable by RPA mission crews and operators, can be shared with the pilot of a manned aircraft, the exploitation component or a ground-based tactical party as directed by joint CONOPS; commander's intent; or service tactics, techniques, and procedures (TTP). In the cases of distributed payload control, CONOPS, flight procedures, and aircraft systems must always provide the crew with sufficient SA to accomplish the mission legally, safely and effectively.

Pilot-monitored, fully autonomous takeoff and landing are proven capabilities of DoD UAS. As technologies mature, automation will evolve to safely execute even more aspects of transitory flight. The "loyal wingman" concept describes autonomy used to increase the mission effectiveness of manned platforms. This FoS may be used for SEAD, AI, and ISR, with the capability for aerial refueling. Modular and autonomous technologies advance the level of flexibility and effectiveness of Group 4 FoS for the JFC. Cooperative engagement will link RPA into aerial formations to simplify enroute transit and enable machine-to-machine links between MUM aircraft.

Group 5 UAS: The Air Force vision for a Group 5 RPA is similar to the Group 4 RPA evolution (see Figure 17), leveraging autonomous, modular, and open architecture technologies. For example, Group 5 RPA acquisition authorities may consider a modular payload design to facilitate tailored capabilities for new theaters or GCC HALE missions. This could be a step toward the NextGen large RPA, which will evolve from high-altitude ISR and battle management command and control (BMC2) to be capable of performing today's manned heavy aircraft missions, potentially with one common core airframe. Filling urgent CCDR needs first, large RPA with SAR/GMTI or advanced SIGINT capabilities will complement existing airframes in ISR missions. Advanced OPA may offer the potential to overcome access challenges (e.g., airspace, host nation) while adding mission flexibility. With proper HSI and the ability to control multiple aircraft, manpower requirements may be reduced during loiter and transit operations due to increased automation and autonomy. These manpower efficiencies are amplified when multiple large payload aircraft are teamed together through loyal wingmen technology under the direction of an aircrew.

The NextGen large RPA will be a multi-mission endurance aircraft, capable of ISR, EW, communications gateway employment, and air mobility operations. This platform will also likely comprise major components of the JALN by providing portions of the high-capacity backbone (HCB) and distribution/access/range extension (DARE) capabilities. These capabilities will enable a large RPA FoS approach through modularity. Appropriate sets of payloads will be modular in nature and use plug-and-play payload bays. Some of the potential payloads include ISR, EA, BMC2, pallet lift capability, or fuel tanks, as required to balance the force. Though the goal is modularity in principle, there is no expectation that any one aircraft will be capable of every mission. Autonomy advances for auto takeoff and land will permit safer integration with civil and military traffic. Loyal wingmen technology will mature such that formations of manned and remotely piloted aircraft could disperse to land at the point of need separately from each other. As technologies mature, ground operations, from taxi through ground refueling and standard pallet loading, will be conducted with only human monitoring of autonomous actions.

Air Force RPA will leverage open architecture control interfaces. Software services developed to the common/universal DoD standard will be reused to the maximum extent practical. Where possible, payloads must be modular in nature to allow for acquisition efficiency while maximizing operational flexibility. Some mission requirements may call for specialized RPA that have characteristics that make them difficult to integrate. Finally, extreme performance parameters, such as ultra-long endurance or hypersonic flight, will demand high levels of autonomy. These systems may require reconsideration of maintenance and logistics support to adequately service the aircraft.

The maturity of the technologies required to support future missions are keys to the success of this vision. As an example, current stealth technology is sufficient to meet today's threats, but stealth technologies that would allow long loiter in high-threat environments require further development. Extremely long-endurance platforms, including high-altitude balloons or large lifting surface aircraft, are under development and could be available in the near-to-mid-term timeframe and may impact CONOPS and vision implementation. The least mature and longest lead technologies are hypersonic systems that rely on propulsion technology and materials that can withstand the extreme heat generated. The prioritization of funding for technology development should be aligned with strategic guidance to look beyond near-term GCC requirements and ensure capabilities support future warfighter needs.

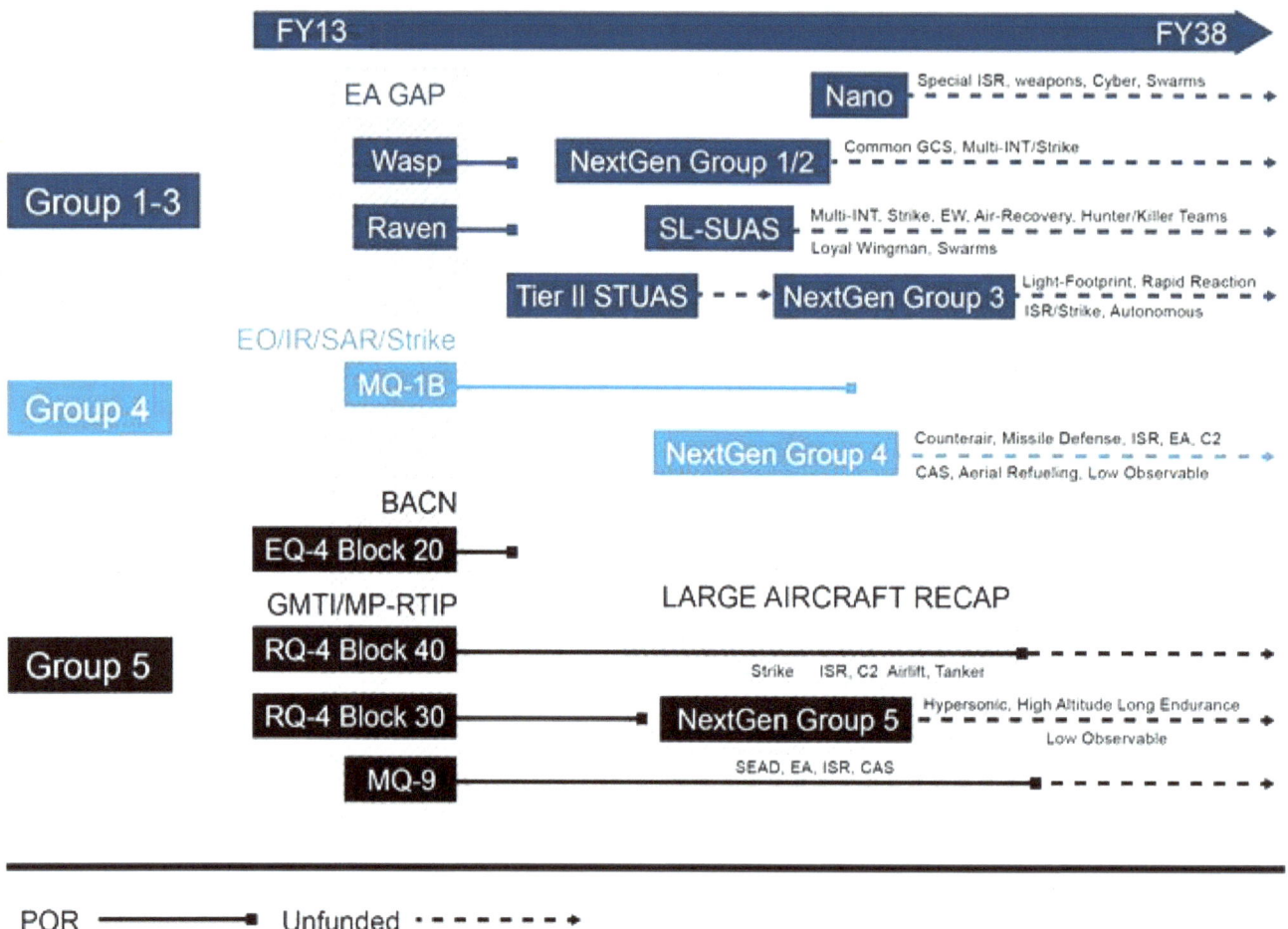

Figure 17: Future Mission Evolution by FoS

3.5   Core Function Master Plan Mission Integration

The intent of this section is to perform a baseline review of Air Force RPA capabilities that leverage the CFMP guidance and applicable studies. Furthermore, this section is intended to broaden awareness and explain contributions that RPA and SUAS capabilities provide and will review, compile, and integrate past and ongoing analysis that could inform senior leaders' decisions. It will include a synthesis of (1) common themes from past studies and (2) findings from efforts currently underway to provide insights based on scenarios derived from current national security guidance.

> *"The Air Force's Service Core Functions...form a reference point for helping the service mold its strategic priorities, risks, and tradeoffs."*
>
> *- Secretary of the Air Force (2008-2013) Michael Donley, 26 February 2009*

The CFMP forms a common framework linking strategic planning and programming to improve what the Air Force brings to the joint fight. In support of this, the CFLI provide agile leadership to help the Air Force achieve the strategic and operational objectives of the National Defense Strategy with projected resources at the lowest possible overall risk.

Table 2: Future Missions versus Core Function

CURRENTLY SUPPORTING	PLANNED	FAR TERM POTENTIAL	
Agile Combat Support Protect	**Command and Control** Global C2	**Air Superiority** Air and Missile Defense Counterair Electronic Warfare Suppression of Enemy Air Defense	**Rapid Global Mobility** Aeromedical Evacuation Airlift Air Refueling
Command and Control Theater C2	**Special Operations** Battlefield Air operations		**Space Superiority** Space Force Enhancement (Launch Detection) Space Force Enhancement (MILSATCOM) Space Force Enhancement (PNT)
Global Integrated ISR Airborne ISR Collection		**Global Integrated ISR** ISR Analysis and PED	
Global Precision Attack Air Interdiction Close Air Support		**Global Precision Attack** Strategic Attack	**Cyberspace Superiority** Cyberspace Offense Cyberspace Defense
Building Partnerships Shape and Communicate		**Personnel Recovery** Humanitarian Assistance Medical Evacuation Search and Rescue	**Special Operations** SOF Mobility
Special Operations SOF ISR SOF Precision Engagement Aviation Foreign Internal Defense			

NOT CURRENTLY SUPPORTING/UNPLANNED		
Agile Combat Support Field Base Support Sustain	**Nuclear Deterrence Ops.** Nuclear Strike Assure, Dissuade, Deter Nuclear Surety	**Space Superiority** Space Control

Table 2 aligns current capabilities to future missions within the Air Force's Core Functions, and the sections that follow expand on many core functions to show potential areas where RPA capabilities could contribute to the strategic environment and the resulting national security challenges facing the United States over the next 25 years.

RPA development must be justified within the CFMP construct, and CFLIs may choose to pursue RPA as an alternative to a manned platform where it makes sense given specific requirements. In some cases, RPA may provide advantages and opportunities. However, some missions are better suited for manned platforms. The decision to pursue RPA as the solution for a specific requirement is at the discretion of the CFLI and must take into account interoperability of data systems, data links, interfaces, waveforms, weaponry, architecture standards and airspace access constraints.

3.5.1 Agile Combat Support

Lead Integrator: Air Force Materiel Command

3.5.1.1 Protect

RPA provide the security forces the ability to deter, detect, delay, deny, assess, mitigate, or neutralize threats against agile combat support assets. It provides the capability to detect impending attacks and decide on protective measures that will deny an adversary the ability to affect operations. Armed RPA provide an enhanced means of controlling and dominating the base security zone and can mitigate attacks. Specifically, RPA can support protection of friendly forces by detecting, identifying, and neutralizing threats (such as IEDs, mortar tubes, and rocket sites), perform convoy escort, armed over-watch, air base defense and homeland defense missions.

Convoy Escort. RPA are well suited for convoy escort and typically provide convoys many more hours of coverage in a single sortie than manned platforms. RPA could be launched and controlled by operators at a remote base or, in the case of SUAS, directly by the mobile ground team. It is essential that good communication between the lead vehicle and the RPA crew is maintained. A spotter RPA may be able stop a convoy in time to avoid an IED or enemy contact. Convoy escort may escalate into a CAS scenario, requiring collaborative targeting. In some situations, if aircrews are appropriately qualified, the RPA may perform the forward air controller–airborne (FAC-A) role.

Armed Overwatch. When an RPA is providing armed overwatch of friendly personnel on the ground, the same considerations apply as when doing convoy escort. Strong communications, extensive knowledge of the rules of engagement, a highly trained crew and a capable RPA are all required.

Airbase Defense. RPA can provide capabilities to routinely patrol and respond in emergency situations to reduce the effectiveness of enemy attacks on, or sabotage of, a base, ensuring the maximum capacity of its facilities is available to U.S. forces. Currently, Air Force security forces use SUAS to provide surveillance over bases and installations. While SUAS, such as Raven, are an excellent small team asset, their endurance and range are limited. Larger assets may be launched to provide extended surveillance or overwatch. In the future, loitering lethal SUAS could be employed in an "airborne minefield" concept to protect against both air and ground attack.

Homeland Security. RPA are capable of supporting a wide range of homeland security operations, including border patrol, maritime and harbor patrol, counterdrug operations, support of disaster rescue and recovery, and other civil support. The expansion of RSO to include components of air domain awareness will enable coordinated interagency operations. Enabling technologies, such as ABSAA and GBSAA, as well as NextGen avionics are critical to ensure the NAS access required for homeland security missions.

3.5.2   Air Superiority

Lead Integrator: Air Combat Command

NextGen RPA will be essential to achieve dominance in the air battle, which, in turn, permits joint forces to operate anywhere at any time without enemy inference. Teamed, manned, and UA will perform air and missile defense, counterair, EW and SEAD missions.

3.5.2.1   *Air and Missile Defense*

In support of the air and missile defense role, RPA will conduct operations that collect and disseminate, in near-real-time, tactical data that can be used to prosecute time-critical targets. Information will be directly down-linked and distributed to all assets conducting defensive operations via broadcast data links, such as the Joint Tactical Information Distribution System. Advances in weapons technology for future RPA will enable kinetic and non-kinetic (e.g., directed energy) weapons to engage and defeat time-critical missile targets. NextGen RPA will be able to provide pre-launch detection, accurate launch point information within seconds of a theater ballistic missile (TBM) launch, and airborne missile tracking and engagement.

Future air and missile defense platforms will directly contribute to theater strategic objectives by attacking emergent high-value targets (leadership targets; TBM; and chemical, biological, radiological, nuclear, and explosive [CBRNE] associated activities). In addition, key attributes include sensor and payload flexibility, the capability for identification of TBM, sensors capable of track and targeting of launched TBMs, autonomous launch detection and tracking capability and the ability to carry weapons capable of destroying TBMs on the ground or intercept after launch.

3.5.2.2   Counterair

RPA will conduct counterair offensive missions against an enemy's capabilities to force the enemy into a defensive posture. This will be part of a joint C2 operation that encompasses various ISR platforms; air-to-air, air-to-ground, and surface-to-air air weapons. Integrated teams of unmanned and manned assets will perform defensive counterair measures designed to detect, identify, intercept, and destroy or negate enemy forces attempting to attack or penetrate the friendly air environment. In addition, RPA will conduct offensive operations to destroy, disrupt, or neutralize enemy aircraft (both manned and unmanned), missiles, launch platforms, and their supporting C2 structures and systems both before and after launch. Concepts like loyal wingman, swarming, and AL-SUAS will be integrated with the mission package to execute attack operations, fighter sweep, escort and SEAD missions.

Key attributes of future counterair platforms are to provide ISR, deception, jamming, or harassment of enemy forces and air defense systems. These capabilities may be used to attack some targets either too dangerous for manned aircraft or where manned aircraft are not present to respond.

3.5.2.3   Electronic Warfare

Improvements in RPA payload technologies will enable RPA to provide electromagnetic, directed energy, or antiradiation weapons to attack personnel, facilities, or equipment. RPA provide a unique advantage in endurance and risk tolerance over the manned platforms of today. An EA-equipped RPA should be capable of long-endurance missions and able to carry small, low-cost, air-launched and expendable options (e.g., Miniature Air Launched Decoy-Jammer). Specialized EA payloads for SUAS platforms should also be considered where SWaP limitations permit.

The goal of EW is to prevent or reduce the enemy's use of the electromagnetic spectrum and protect friendly personnel, facilities and equipment. In support of this mission, RPA should be equipped to search, intercept, rapidly identify threats and locate sources of radiated electromagnetic energy.

Key attributes of future EW platforms include sensor payload flexibility that allows for specialized EW packages to provide selective jamming, spoofing or EA capability. The vision for EA capability is the integration of mature EA payloads on RPA. The components must be modular in nature to allow the use of interchangeable payloads that meet the threat and are reprogrammable to rapidly adapt to new threats. These platforms and payloads must be network-capable to share information with other EA systems, both manned and unmanned. RPA equipped with EW payloads would provide both self-defense and offensive capabilities for EA operations, as directed by the combined force air component commander. The platform must include countermeasures for survivability, such as towed decoys, missile warning systems, active jammers, and threat alerts while also maintaining sufficient speed, altitude ranges, and maneuverability. The aircraft must be designed with electrical hardening and sufficient power for EW systems. The platform and associated EW accoutrements must enable flexible communications capable of importing target or threat information from other assets. Future antenna and laser designs should enable platforms to create effects independently or by teaming with other manned or unmanned platforms.

3.5.2.4   Suppression of Enemy Air Defense/Destruction of Enemy Air Defense

RPA and SUAS offer potentially expanded SEAD and destruction of enemy air defense abilities to disrupt or destroy air defense targets, such as enemy C2 facilities, radar sites and launchers. Teaming with manned platforms offers flexibility in approaches to deny the adversary's ability to perform air defense functions. RPA must be capable of operating in contested and A2/AD environments. Technologies, like AL-SUAS, provide mission-tailorable modular payload packages that can be used to detect, jam, neutralize or destroy enemy air defenses. Swarming SUAS can be used to saturate enemy defense systems or divert attention through decoying. In addition, RPA and lethal or armed SUAS sent in advance of manned platforms may increase rates of success when confronting a large enemy force. Strategically pre-positioned, lethal or perching network attack micro SUAS may be used to neutralize air defenses without prior detection.

Key attributes for SEAD are similar to the key attributes for CAS and interdiction, requiring operations in contested and A2/AD environments in conjunction with other manned assets. These attributes include external capability for larger weapons (up to 2,000 pounds), high subsonic speed and significant mission range (1,000 nautical miles), sensors with ability to detect components of enemy IADS in all weather and mixed terrain, and multiship cooperative control approaches for searching and engagement. The platform must be capable of sufficient speed, altitude, and maneuverability and possess countermeasures, such as towed decoys, missile warning systems and active jamming equipment.

3.5.3   Command and Control

Lead Integrator: Air Combat Command

The Air Force will use large platforms to provide information to organizations and decision makers performing C2, both globally and in support of operations in the assigned theater. Multi-sensor RPA can be used as surrogate sensors attached to capital surveillance BMC2 weapon systems. They could extend surveillance volumes, act as gap fillers, detach to cover specific areas of interest and add better fidelity information. HALE RPA with communications payloads will provide airborne relay primarily over rugged, mountainous, or urban terrain, where other communications options are limited or to decrease dependence on satellite connectivity. On the battlefield, RPA will provide persistent long-range communications relay or act as a gateway manager of multiple communications to enhance C2 connectivity and span of control.

Key RPA attributes for future C2 platforms include sensor payload flexibility with capabilities for specialized communication radios and antenna. Employment methods must be compatible with multiple aircraft integration to distribute or share connectivity with other C2 nodes/platforms, decision makers, aircraft, and maritime/ground forces. Developments in phased array antennas and smart antennas that combine reception of multiple signal types could offer an alternative to traditional dish antennas; however, they require tradeoffs in SWaP. Continued technology advances, such as multifocused and super-cooled antenna systems, would permit multiple users to receive information and not rely on point-to-point systems and subsequent relaying of data via other communications systems to local users.

3.5.4   Global Integrated ISR

Lead Integrator: Air Combat Command

Today the primary mission of RPA is to conduct globally integrated ISR as an airborne ISR collection platform and to support ISR analysis and PED.

Key RPA attributes for global integrated ISR (GIISR) apply to various subsets of this core function: airborne ISR collection, battle damage assessment (BDA) and SCAR. These key attributes include sufficient speed, altitude, and maneuverability to enable survivability in contested environments and A2/AD while also including countermeasures, such as towed decoys, missile warning systems, and active jamming equipment. To maintain SA, threat alerts and support for night, weather, terrain, and culture awareness must also be included. The platform must be able to support multiple objectives or engagements and, therefore, requires significant mission range and persistence. Flexibility in communications (to import target or threat information from other assets), weapons (e.g., carry a full complement of SEAD weapons), and sensors (e.g., EO/IR, FMV, SAR, SIGINT, light detection and ranging [LIDAR], wide area search [WAS], foliage penetrating [FOPEN], additional pods) is critical. Finally, employment methods must be compatible with multiple aircraft to distribute or share communications to enable teaming with manned or unmanned vehicles for searching and engagement.

> *"Information is a strategic asset. Without it, the organizations and the operations they undertake have little chance of success."*
>
> *– John G. Grimes, DoD CIO*

3.5.4.1   Airborne ISR Collection

RPA will provide ISR data from a variety of sensors over an LOS tactical common data link or derivative, as well as by BLOS links. Other sensors, such as weather sensors, can be employed as mission-specific payloads in a plug-and-play mission kit concept. ISR tasking and missions will be blended with other mission assignments depending upon mission priorities as tasked.

R&D in smaller, lighter, and more sensitive CBRNE sensor packages will increase the capability of UAS. Compact, active multispectral chemical sensors will enable the remote detection of chemicals associated with weapons.

Future UAS must address operating in CBRNE environments. For example, UAS subsystems will need to be survivable and able to continue operations following a high-altitude electromagnetic pulse event from a nuclear detonation.

3.5.4.2   Battle Damage Assessment

RPA are uniquely designed for these missions due to their long loiter times, onboard use of sensors, and integration with PED facilities. RPA BDA configurations must be considered during planning. For example, type of sensor, camera fidelity and zoom levels are important to consider. Gateways to Link 16 and multifunction advanced data link will enable RPA to directly send BDA to ground forces and also to strike platforms that are still in the target area.

3.5.5   Global Precision Attack

Lead Integrator: Air Combat Command

Building on today's capability to find, fix, and finish, future RPA, including potential unmanned versions of long-range strike platforms, must possess the ability to detect, locate, identify and engage a wide variety of targets anywhere on the globe while limiting collateral damage. RPA will conduct SCAR missions to detect targets and coordinate or perform attack or reconnaissance on those targets. RPA will leverage their effective combat radius, sensor suites, and weapons to perform SCAR.

Key RPA attributes of global precision attack (GPA) are similar to GIISR core function key attributes. These similarities of key attributes include sufficient speed, altitude, and maneuverability to enable survivability in contested environments while also including countermeasures, such as towed decoys, missile warning systems, and active jamming equipment. To maintain SA, threat alerts and support for night, weather, terrain and culture awareness must also be included. The platform must be able to support multiple objectives or engagements and, therefore, requires significant mission range and persistence. Flexibility in communications (to import target or threat information from other assets), weapons, and sensors (e.g., EO/IR, FMV, SAR, SIGINT, LIDAR, WAS, FOPEN) is another key attribute. Finally, employment methods must be compatible with multiple aircraft to distribute or share communications and to enable teaming with manned or unmanned vehicles for search and engagement.

In addition, specific platform requirements include the capability for low and slow flying as well as high supersonic speed and mission ranges of 1,000 nautical miles. Flexible and precise weapons are a key attribute for the mission set and include carrying a full complement of ordnance, both internally for stealth and up to 2,000 pounds externally in a non-A2/AD environment.

3.5.5.1   Air Interdiction

RPA leverage persistent loiter, flexible sensor capabilities, and organic weapons to hunt and kill both preplanned and time-sensitive targets along LOC or in areas of known or suspected enemy activity in support of commanders' needs. Imported data from WAS platforms, fused to a COP feed, provide SA and augment initial target cueing. NextGen RPA will be employed stand alone or in concert with other assets, with machine-to-machine data links, target designators and markers, and secure voice communications in "buddy" or coordinated attacks when tactically expedient.

3.5.5.2   Close Air Support

Some CAS functions have been performed by MQ-1 Predator and MQ-9 Reaper aircraft in current conflicts. The long-endurance and payload capabilities of RPA have enabled extended time over target, which allows for enhanced SA and direct support to warfighters. The NextGen RPA pilots will conduct persistent FAC-A as well as weapons delivery missions using a combination of organic and offboard sensors and precision ordnance. It will possess the capability to provide rapid on-call precision weapons to accurately employ supporting fires when needed as well as coordinating joint fires. Supporting technologies for the full CAS mission are being developed today through programs, such as the Defense Advanced Research Projects Agency (DARPA) Precision CAS optionally piloted A-10 demonstration.

3.5.5.3   Strategic Attack

RPA can directly contribute to theater strategic objectives by attacking emergent high-value targets (such as leadership targets, TBMs and CBRNE-associated activities) in conjunction with long-range strike assets.

3.5.6   Nuclear Deterrence Operations
Lead Integrator: Air Force Global Strike Command

3.5.6.1   Nuclear Strike

Certain missions, such as nuclear strike, may not be technically feasible unless safeguards are developed and even then may not be considered for UAS operations. On that issue in particular, Headquarters Air staff will be integral to the development of UAS roles in the nuclear enterprise and vetting through the Joint Staff and GCCs. Ethical discussions and policy decisions must take place in the near term to guide the development of future UAS capabilities, rather than allowing the development to take its own path apart from this critical guidance.

3.5.7   Personnel Recovery
Lead Integrator: Air Combat Command

RPA support PR missions today through capabilities to identify and communicate with isolated U.S. military and civilians and assist their recovery to friendly control. One example of this support occurred at the start of OEF when a Predator aircraft performed a key role in recovering imprisoned U.S. citizens who were missionaries in Afghanistan. RPA currently conduct reconnaissance for ingress/egress routes and extraction points. The recovery locations will generally be remote with very limited landing access in mixed terrain and culture (covering urban to rural). These locations may exist in contested threat environments, and the personnel may be hidden or under fire. NextGen RPA transport (CQ-X) may deliver supplies (e.g., survival gear, radios, weapons), conduct radio relay, or land to recover isolated personnel. In addition, MQ-X or strike RPA will provide armed escort or SEAD support to PR missions.

Key attributes for future PR platforms include flexibility (enabling adaptation to an ever-changing tactical or operational situation in a dynamic environment and fluctuating logistics scheme) and precision to maneuver in a timely manner to a specific location. The platform requires survivability attributes for operations in environments where the UA may experience hostile fire, adverse weather conditions, challenging terrain, and high elevations or remote locations. Future platforms must be able to team with other manned or unmanned vehicles. Additional key attributes that apply to PR platforms are included in the Medical Evacuation/Casualty Evacuation (MEDEVAC/CASEVAC) and CSAR sections.

3.5.7.1   Combat Search and Rescue

The NextGen RPA will augment combat search and rescue missions with the capability to provide constant survivor status and geo-location data through its extended loiter capability. The NextGen RPA will also support the combat search and rescue mission by detecting and identifying potential combat threats to the recovery assets and survivor, with the means to neutralize threats whenever they appear. The combination of operator SA, persistence, sensors, and communications would make the RPA suitable for on-scene commander responsibilities.

The key attributes for combat search and rescue are similar to those for GIISR and GPA, with specific requirements for robust communications and networking capabilities that enable teaming with manned and unmanned vehicles. Future platforms should include sufficient speed, altitude, maneuverability and defensive systems to enable survivability in contested environments while also including countermeasures, such as towed decoys, missile warning systems and active jamming equipment. In addition, SA enhancements, including threat alerts, night vision aids, weather, and terrain and culture awareness, must also be included. The platform must be able to support multiple objectives or engagements and, therefore, requires significant mission range and persistence. Flexibility in communications (to import target or threat information from other assets), weapons and sensors (e.g., EO/IR, FMV, SAR, SIGINT, LIDAR, WAS, FOPEN) is also critical to mission success.

Key attributes specific to the combat search and rescue mission of the PR core function include platform requirements for low and slow flying as well as mission ranges of 1,000 nautical miles.

3.5.7.2   Medical Evacuation/Casualty Evacuation

The primary differences between MEDEVAC and casualty evacuation (CASEVAC) are the assets conducting the missions. MEDEVAC is normally conducted in an uncontested environment by dedicated, unarmed platforms that are staffed and equipped to medically support a patient en route to a medical facility. CASEVAC is normally conducted in a contested environment by an armed battlefield asset, from the point of injury to a medical facility, by a platform that may or may not have medical personnel and equipment on board. During the Joint Expeditionary Force Experiment, Network-Tactical (NET-T) capable ROVER and surrogate RPA linked real-time patient vital information across the network from the simulated injured person to the main base medical team. As technologies advance, MEDEVAC may be performed by either OPA or RPA with autonomous, robotic and medical personnel performing patient care on board. When combined with future telesurgery capabilities, medical data connectivity and advanced robotics integrated into future RPAs will maximize their success as first responders in areas inaccessible to manned recovery aircraft.

With additional potential capabilities, such as vertical takeoff and landing (VTOL) or short takeoff and landing, the forward unit commander could evacuate wounded personnel from a forward operating base or transit facility back to the main operations base. The capability to precisely deliver time-sensitive/mission-critical sustainment to a forward area underscores the use of a theater RPA especially during a CBRNE environment. In this case, use of an RPA would eliminate airborne aircrew exposure to potential CBRNE hazards.

Additional key attributes specific to MEDEVAC and CASEVAC include a capacity range from one to eight troops, comprehensive ground SA, defensive capabilities, flexible payload capacity for personnel and medical facilities/supplies, and adaptability to respond to redirection or routing to execute a MEDEVAC or CASEVAC mission while conducting resupply operations as a cargo asset. The platform must be capable of very low speeds and altitudes as well as the ability to conduct a VTOL. Future platforms should consider autonomy and have sufficient power capacity for loading and unloading personnel and equipment in remote locations.

3.5.7.3   *Humanitarian Assistance*

In the past, Predator and Global Hawk have supported humanitarian assistance missions after natural and manmade disasters by tracking wildfires, flood impact areas, nuclear reactor damage, and locations of isolated personnel. This role will increase for National Guard units especially with integration of new platform capabilities. Potentially CQ-X may deliver supplies (e.g., food, water, blankets) or land to recover personnel.

3.5.8   Rapid Global Mobility

Lead Integrator: Air Mobility Command

3.5.8.1   *Airlift*

Airlift is the transportation of personnel and materiel through the air, which can be applied across the entire ROMO to achieve or support objectives and can achieve tactical benefit through strategic effects. Airlift provides rapid and flexible mobility options that allow military forces as well as national and international governmental agencies to respond to and operate in a wider variety of circumstances and timeframes.

There is potential for OPA airlift platforms to provide a near-term capability to augment airdrop/resupply cargo missions using existing mobility aircraft. As an RPA designed to support the transport mission, CQ-X can provide an alternate delivery option to joint/coalition forces distributed across the area of operations for routine, around-the-clock, and time-sensitive logistics support to widely dispersed units within the joint task force. CQ-X will reduce the threat to manned cargo airlift in high threat environments and can augment or replace ground vehicles delivering high-priority parts, medical supplies, or other equipment to threatened areas or to remote units with no vehicle access. This unmanned cargo capability could also enable resupply from sea-basing assets located offshore, provide support for SOF, and release manned cargo aviation assets for more demanding missions.

The addition of CQ-X will significantly enhance joint force operational flexibility and response by supplementing the capabilities of manned airlift assets. CQ-X can be uniquely suited to extend crew duty day, risk flying into reduced visibility when resupply is critical, deliver cargo into dangerous situations, and reduce risk to manned assets.

A theater-level CQ-X, integrated with tactical-level unmanned and manned cargo aircraft, improves responsiveness and agility. A tactical-level CQ-X (VTOL) would be organically owned by the brigade or battalion and moved closer to the fight. This could release current tactical aircraft from sustainment to conduct higher priority missions (e.g., fire support, maneuver). The theater CQ-X would be capable of transporting an internal cargo payload (approximately three to five pallets) and be remotely piloted from a location within the theater. The vehicle would be equipped with onboard automated material-handling equipment for quick and simple ground onload and offload capability. These capabilities and parameters would give this medium-sized RPA a true operational/theater-wide ability to supplement manned airlift in sustaining the joint force. In addition, the increased size and payload potential of this type of RPA would be conducive to hosting multi-mission packages.

A CQ-X could also be developed and employed to sustain the joint/coalition force from strategic distances. This air vehicle would be capable of airlifting a significant payload of equipment and sustainment cargo over strategic distances (e.g., CONUS-AOR).

Key attributes for the CQ-X airlift mission include flexibility and adaptability to support the ever-changing requirements of the overall logistics scheme of a tactical or operational situation within the area of operations. The platform must be survivable in threat conditions and in adverse weather conditions. CQ-X must be able to conduct operations independently and also team with manned or unmanned logistics assets.

3.5.8.2   *Air Refueling*

Air refueling (AR) is an integral part of global air mobility and brings added capability to combat, combat support and air mobility for all airpower operations. AR enhances the unique qualities of airpower across the ROMO. Furthermore, AR significantly expands force options available to a commander by increasing the range, payload, loiter time and flexibility of other aircraft.

Basic missions include inter-theater and intra-theater AR support: global attack support, air bridge support, deployment support, theater support to combat air forces and special operations support. Inter-theater AR supports the long-range movement of combat and combat support aircraft between the CONUS and a theater, between theaters, or between theaters and joint operations areas. Inter-theater AR air bridge operations also support execution of global strike and long-range airlift missions. AR enables deploying aircraft to fly nonstop to their destinations, reducing closure time. Intra-theater AR supports operations within a CCDR's AOR by extending the range, payload, and endurance of combat and combat support assets.

The Air Force Research Laboratory (AFRL), in cooperation with several industry partners, has developed and tested an automated air refueling (AAR) capability with a manned KC-135 and a simulated unmanned combat aircraft. In addition, DARPA conducted flight demonstrations with two RQ-4s, equipped with AR transmitter and receiver equipment, demonstrating rendezvous through stabilized pre-contact position as a first step for the future development of AAR. This emerging capability would allow manned airborne tanker aircraft to air refuel remotely piloted combat and combat support aircraft, extending their range or loiter time. AAR technology could also be used in AR for manned aircraft as backup to mitigate risk (e.g., low visibility). In addition, the technology could be applied to a future remotely piloted tanker aircraft to air refuel unmanned or manned combat and combat support aircraft. With no crew to limit flight duty period and large quantities of fuel onboard, tanker sortie range and loiter time would be significantly improved.

Key attributes for a future RPA for air refueling (KCQ-AR) could include use of AAR for tanker/receiver rendezvous, track procedures, wing formation, pre-contact and contact positioning. Additional autonomy is required for tactical profiles and advanced navigation systems for enroute accuracy and access to global airspace. The KCQ-AR needs a significant combat radius and low-to-high airspeed ranges to accommodate AR of other Air Force, sister services, allied and coalition aircraft. The C2 of a KCQ-AR requires flexibility and adaptability to the ever-changing operational and tactical requirements. Additional key attributes for enhanced survivability include reduced radar signature, communications, and data link relay systems capable of autonomous defensive systems against IR and RF threats.

Ultimately, a KCQ-AR should have the same key attributes and mission requirements as a manned airborne tanker aircraft except for those specific requirements that support a tanker aircraft aircrew.

3.5.9   Space Superiority

Lead Integrator: Air Force Space Command

3.5.9.1   *Space Control*

Special-mission RPA, capable of near-space operations, are being conceptualized to support U.S. dominance in space. This may include hypersonic near-space "mother ship" RPA that deliver multiple SUAS to provide a strategically significant number of lethal, EA, or cyberattack capabilities within minutes. In addition, many RPA will be capable of integrating sensors that will support space force enhancement through launch detection. In the near term, this may be the same sensor/UA configuration used to support air and missile defense missions.

3.5.10 Building Partnerships

Lead Integrator: Air Education and Training Command

3.5.10.1   *Shape and Communicate*

Interoperable RPA will enhance our nation's ability to conduct activities with other nations' militaries, which is vital to success in future coalition conflicts and helps to improve their population's perception of the U.S. military. The opportunity exists, especially with SUAS, to improve their capabilities through foreign military sales of UA as well as training and operator engagement activities. Furthermore, modular UA will quickly integrate payloads or aircraft components manufactured by U.S. partner nations to enhance the success and acceptability of U.S. UA. An example of a modular interface is the Air Force–developed UAI, to be implemented on DoD UA. It is currently recommended as a North Atlantic Treaty Organization (NATO) standard. Through the implementation of this standard, all NATO munitions will eventually be compatible with all weapons platforms, enabling NextGen UA to be more readily exportable to other NATO countries. This will also expand U.S. ability to train and deploy with partner nations, strengthening theater engagement and partnership alliances.

3.5.11 Cyberspace Superiority

Lead Integrator: Air Force Space Command

Dominance in cyberspace by the United States has the potential to be greatly enhanced by RPA, especially in geographically separated theaters. Cyberspace is a fluid domain, much like the air or sea, in that it is not constrained by geography. It is almost impossible to control or contain the reach of cyber influence. Superiority in this domain requires the Air Force to identify and neutralize threats without being able to fully control the environment. The Air Force needs to exploit the domain in the vicinity of the enemy as for all other airpower domains. The attributes of RPA offer both significant advantages and vulnerabilities in achieving cyberspace superiority.

3.5.11.1   Cyberspace Offense

Through cyberspace warfare, RPA can achieve effects globally. Current cyberattacks are limited by the ability to infiltrate enemy networks through natural choke points. Significant access to enemy networks from outside the geographic boundaries relies on a limited number of fiber optic entry points and SATCOM ground terminals. The enemy can focus efforts at those locations to try to block entry of DoD cyberwarfare actions. Even unintentional actions could prevent the DoD from operating on other nations' networks. Recently a major fiber optic cable connecting Western Australia to the world was accidentally cut underwater. This event isolated Western Australia's networks from everyone, both allies and enemies. The Air Force could be limited in the ability to conduct operations if physical geographic boundaries of an enemy nation were denied. Larger NextGen RPA will be able to employ cyberspace operations through collecting information on the enemy network and exploiting vulnerabilities. SUAS, particularly micro systems, could perch near network inject locations and perform cyberattacks, synchronized with other space and air domain operations.

The DoD also has the opportunity to surreptitiously "borrow" enemy computing power and data transmissions without altering enemy systems, similar to commercial/educational projects that share processing workloads across several volunteer computers to multiply the real-time computer processing available. Complex processing could be done with LOS networks without SATCOM delays or the need to wait for RSO processing.

3.5.12   Special Operations

Lead Integrator: Air Force Special Operations Command

Specialized airpower operations are regularly conducted by RPA in hostile, denied, or politically sensitive environments. This includes a variety of UA, some with covert, clandestine, or low-observable/low-detectable capabilities. MQ-1/9 and future RPA work in conjunction with the covert air and ground systems and teams to conduct battlefield air operations and SOF ISR. Covert RPA and SUAS must be hard to detect visually, thermally, on radar and acoustically because knowledge of their presence could compromise the operation.

3.5.12.1   Battlefield Air Operations and SOF Precision Engagement

Weaponized RPA, such as the MQ-1/9, are integral to SOF team operations. Precise RPA weapons employment perfectly synchronized with other air and ground actions multiplies the surprise and combat effectiveness of small covert forces. Furthermore, lethal AL-SUAS have been employed through launch from MQ-1/9 and guided to impact and destroy an objective using BLOS links. Kinetic, EA and cyberspace attacks conducted through RPA will be critical to future SOF battlefield air operations.

3.5.12.2   Special Operations Forces ISR

Precise dedicated ISR is leveraged by SOF to give it an asymmetric advantage for its operations. Future platforms should include sensor payload flexibility with the capability for specialized SOF communications and use all source-fused data from FMV, SIGINT, radar and other sensors to provide direct threat warning and enhanced SA to AFSOC aircrews. Remote terminals, such as ROVER, will soon be NET-T enabled, which will greatly expand the shared SA of the aircrews, ground teams, and for some operations, the senior decision makers. This capability also applies to their actions to support humanitarian relief. Platforms should include multi-aircraft compatibility to distribute or share communications with other SOF aircraft or SOF ground forces.

3.5.12.3   *SOF Mobility*

Future CQ-X must be capable of delivering specialized on-demand cargo to small groups of troops generally located in remote areas. The delivery mechanisms must include conventional ground-based cargo delivery and precision airdrop. For conventional cargo delivery, cargo unloading must be rapid, highly autonomous, and capable of communication with unmanned ground and maritime platforms. A precision airdrop limits movement or exposure of receiving units while receiving critical supplies behind enemy lines. One critical consideration for air dropped delivery containers is their ability to be easily broken down, hidden, or destroyed. SOF drop/delivery locations will generally be remote with limited access to basing or improved runways, and in most cases no runway at all. Small CQ-X offer a force multiplier when teamed with conventional rotary wing assets and should be capable of operating in a hostile environment to deliver critical supplies.

Future platform desired payload capacity will range from 200 pounds of basic survival supplies up to three conventional pallets at a total payload weight of 18,000 pounds. The aircraft must consider flexible payload capacity for pallets and odd packages with internal carry or external pod delivery and use easily accessible loading and unloading methods. As the cargo delivery locations may be in contested threat environments, the platform must include appropriate countermeasures, speed, and maneuverability as well as comprehensive SA on the ground. Future aircraft must be capable of very short takeoff/landing using unimproved runways or low-altitude precision airdrop over remote locations. Considerations for autonomous cargo handling for remote ground operations and appropriate power for sensors, cargo load/unload, and airdrop systems must be considered.

3.6   DOTMLPF Considerations

In the development of the RPA Vector, the unique characteristics attributable to RPA and the potential missions in which RPA could be employed to enhance combat effectiveness were considered. Though not formally prioritized, the identified key attributes were viewed through the lens of DOTMLPF to articulate the Air Force decisions required to achieve the requisite capabilities. Because the RPA Vector spans a wide range of systems across potential missions over a 25-year period, the solutions are assembled as a portfolio of capability milestones over time and provide the initial steps for future CBA and analysis.

Materiel solutions are insufficient to achieve critical capabilities without corresponding DOTMLPF actions. Some of these actions may take as long to accomplish as the technology development for the materiel segment of the capability.

Figure 18 highlights some of the key actions that will need to be synchronized to achieve the vision and enabling concepts described in Sections 3 and 4. The DOTMLPF actions are shown across a relative time scale with near-, mid- and long-term sets of capability goals. Specifically, this aligns the future mission evolution by FoS with the key processes and strategic guidance. Thorough review by the key process owners will be critical to meet the JOAC challenges that only fully integrated domain capabilities can accomplish.

In the near term, the majority of the materiel solutions are selective upgrades to existing systems that align with the vision. Current investment in training simulators will streamline initial training. Continuation and virtual training must be expanded for RPA and SUAS in Air Force and joint training exercises in advance of combat deployments. One of the keys to success is the development of TTP to better integrate C2 and ISR with manned and remotely piloted systems.

The total force is investing in facilities to meet demands of its RPA units and depots. Investments in infrastructure and facilities will continue over the next 25 years. New organizations and unique future systems may require new types of facilities as well that must be defined as these capabilities approach fielding.

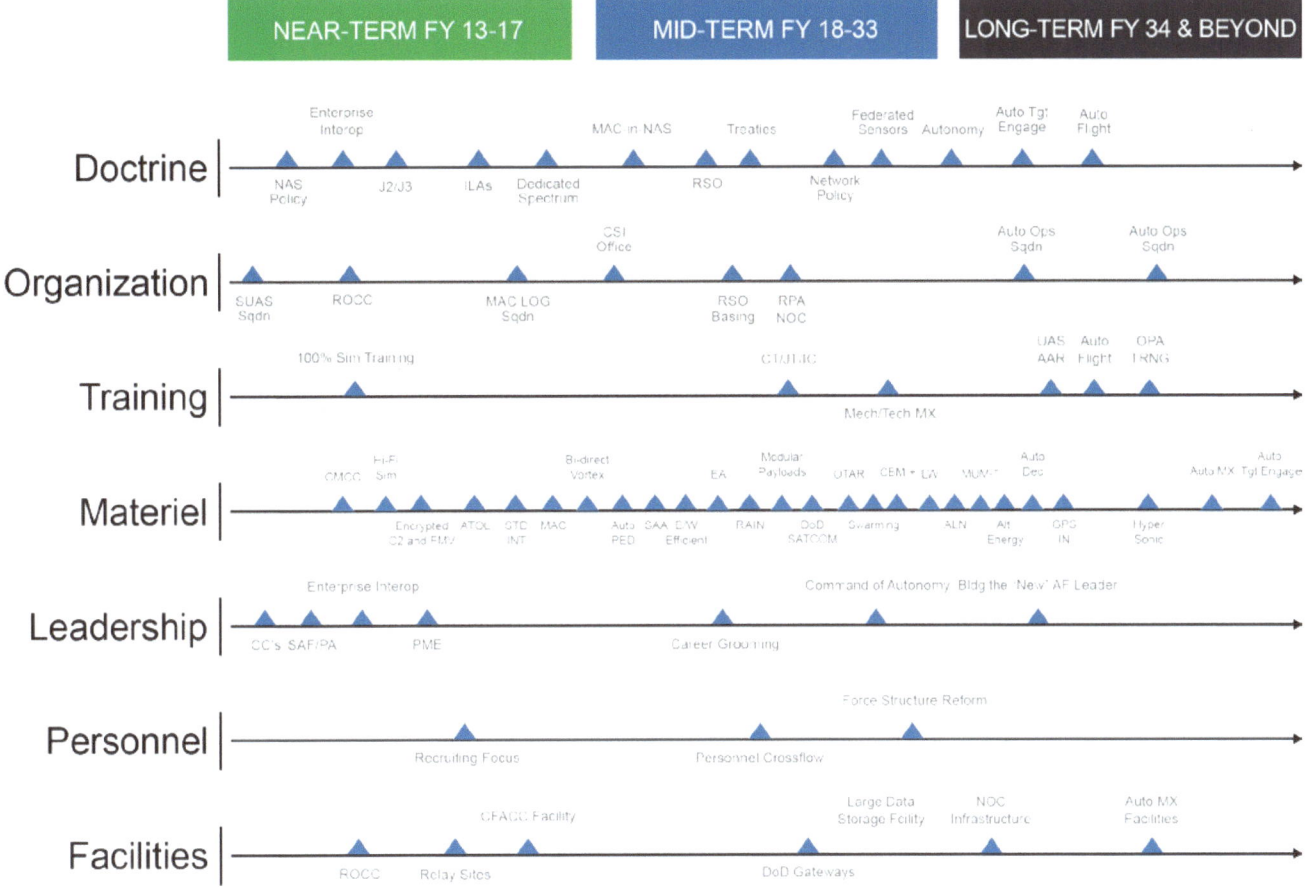

Figure 18: DOTMLPF Synchronization

The mid-term should build on the capabilities and lessons learned in the near term, and the most significant advances will align DOTMLPF for integration of NextGen RPA. While some of the technologies developed in the near term will be integrated in the current platforms, others will be initially fielded with NextGen RPA.

By mid-term, the force structure should be at a sustainable level and capabilities should align to meet the JOAC environments. The Vector anticipates that this will require a renewed focus on personnel and leadership to select, train, and retain the right skills and experiences.

The depiction of long-term synchronization makes many assumptions, not least of which is to assume that the capabilities in the near and mid-term are fully integrated and key advanced technologies have matured. If the fiscal environment allows and process owners have been successful in prioritizing and synchronizing the DOTMLPF actions, the RPA vision and enabling concepts will be fully realized in the long term.

4.   ENABLING CONCEPTS

The previous section, Strategic Enterprise Vision, set the foundation for the vision for RPA at a very high level. This section, Enabling Concepts, will build on those key tenets and further discuss in greater detail the operational impacts, S&T, and R&D initiatives that are ongoing or are still required to support the envisioned capability growth. The discussed enabling concepts address communications issues; interoperability and C2; autonomy; airspace integration; sensors, payloads, and weapons; training and human interfaces; and improved platform capabilities. The areas discussed in this chapter attempt to baseline our current capabilities and highlight advances needed to meet the future challenges.

> *"Change is ever present. Change brings both stresses and opportunities. Agility is the ability to successfully cope with change. Change for change's sake is not agility. Agility implies effectiveness."*

4.1   Agile, Secure, Efficient and Robust Communications

Communications that enable RPA operations to cover all domains (air, space, cyberspace, and terrestrial/maritime) and improvements must address the challenges across the domains. There are a number of enabling capabilities required to achieve the vision for agile, secure, efficient, and robust communications. These include commercial and military satellites, improved spectrum management, bandwidth efficiencies, methods to ensure communications and aerial network enablers. The current RPA architecture is costly and inefficient, and the security challenges require active management of the communications links, encryption, and improved means to reduce the amount of data transmitted through compression and onboard processing. Realizing this vision under the existing stove-piped framework may be an unachievable goal.

> *"Modern armed forces cannot conduct high-tempo, effective operations without reliable information and communication networks and assured access to cyberspace and space. Today, space systems and their supporting infrastructure face a range of threats that may degrade, disrupt, or destroy assets. Accordingly, DoD will continue to work with domestic and international allies and partners and invest in advanced capabilities to defend its networks, operational capability, and resiliency in cyberspace and space."*
>
> *– Sustaining U.S. Global Leadership: Priorities for 21st-Century Defense*

4.1.1   Space Layer Communications

4.1.1.1   Commercial SATCOM

While many of today's RPA make extensive use of commercial satellite communications (COMSATCOM), and it has proven to be largely effective in meeting mission needs in today's fight, this service has drawbacks predominately associated with availability, capacity and coverage. First and foremost, COMSATCOM is an open commodity wherein DoD competes with numerous other communications customers (e.g., television, international telephone, data). Also, COMSATCOM transponders are typically sized for the community they intend to support, most commonly 36 MHz. While that transponder size is sufficient for Predator/Reaper (approximately 10 MHz), it is less than adequate to support Global Hawk's full throughput needs, which can exceed 100 MHz. Coverage has also historically been a concern when using COMSATCOM. Transponder beams are designed to support commercial industry and may or may not be where they are needed for RPA operations, leaving coverage gaps in regions of critical interest.

Given the cost of new SATCOM terminal development in a fiscally constrained environment, it is anticipated that all needs of MQ-1 and the vast majority of MQ-9 SATCOM will be met by commercial Ku-band SATCOM for the foreseeable future.

Finally, while costs vary with each lease, OCONUS COMSATCOM bandwidth may exceed $80,000 per MHz per year (as of 2012). That cost is highly variable based on operating locations and current market demand. If all MQ-1/MQ-9 CAPs remained on COMSATCOM, the annual recurring cost would be more than $75 million. As a result of the large recurring cost of COMSATCOM leases and their vulnerabilities and limitations, the Air Force must look at other enterprise-wide alternatives and planning methods to satisfy RPA requirements that are more efficient and cost effective in the long term and support coverage and protection requirements where needed. Programs will also need to make more efficient use of bandwidth by processing data onboard the platforms or only offboarding critical data pertinent to other missions (while the remainder of the data will be downloaded once the platform lands).

4.1.1.2   Wideband Global SATCOM

Wideband Global SATCOM (WGS) is the primary wideband military satellite communications (MILSATCOM) solution to support military wideband communications and data needs for the next 15 years and beyond. The WGS constellation is currently planned as an eight-satellite geosynchronous constellation (four on orbit as of 2QFY12). The first three satellite vehicles (SV) Block 1 WGS satellites provide up to 125 MHz maximum bandwidth per user, and SVs 4–8 (Block 2 WGS) satellites will provide up to 400 MHz maximum bandwidth for up to two users per satellite. However, given that WGS cannot "auto track" RPA missions, it is better suited to MQ-9 missions than longer-range (long-track) systems, given that MQ-9 may operate within one antenna beam footprint for the majority of its mission.

Global Hawk will not transition to WGS in the near term and instead will rely on commercial Ku-band SATCOM. Efforts are underway to study follow-on SATCOM solutions to meet its long-track, high-bandwidth requirements. Global Hawk may also add wideband International Maritime Satellite Organization (INMARSAT) capability to enable C2, with voice communications, as a redundant BLOS C2 capability.

As the Air Force begins to look at requirements and capabilities across the ISR Enterprise, all SATCOM solutions must be considered. To ensure BLOS C2 in the future, RPA must increasingly consider, based on mission, cost, schedule, and risk, incorporating protected MILSATCOM.

4.1.1.3   Protected MILSATCOM

In many instances protection of critical communication paths and the security of the information flowing through them is vital to national security interests. Satellite, air and terrestrial systems can encounter a number of threats, including jamming, interference, direction-finding, interception, intrusion, physical attack, ionosphere scintillation and other effects (e.g., nuclear detonation). In the future, C2 and, to a lesser extent, wideband payloads will be available via "protected" communications, such as the advanced extremely high-frequency (AEHF) constellation. Based on the electromagnetic spectrum they operate in and the capabilities built into AEHF satellites, they can provide global, highly secure, protected, survivable communications for joint forces. However, because AEHF is a much more complex, "processed" satellite system, its terminals are more complex and costly. At the very least, additional antenna are often required, making protected MILSATCOM an expensive venture today;

platforms may also be constrained by SWaP limitations of adding new RF equipment. Research work is ongoing to reduce the cost of protected MILSATCOM terminals and make them affordable to larger numbers of platforms, including RPA. Future RPA platforms, especially those participating in contested or A2/AD environments should consider protected MILSATCOM as a requirement for assured C2 and mission data.

4.1.1.4 *Inclined Orbit SATCOM*

Inclined orbit SATCOM tracking capability is in use by RQ-4. It has also been demonstrated with MQ-1/9 and is currently undergoing evaluation for operational use. Inclined orbit satellites are typically older, degrading satellites that vendors push into inclined orbits for their remaining operational years, allowing the vendor to put a new satellite in its orbital slot. Inclined orbit satellites typically have significant vacant bandwidth capacity available and at a much lower price per MHz than regular satellites. AFSPC, in cooperation with ACC and Air Force Materiel Command (AFMC), should explore opportunities with Defense Information Systems Agency and industry for reducing SATCOM costs as part of overall SATCOM planning efforts.

4.1.1.5 *Spectrum Management*

Available RF spectrum, just like fuel or power, is an essential enabler for RPA operations. The portions of the RF spectrum having favorable propagation or data-carrying properties are a highly valuable commercial commodity. Government authorities worldwide continue to see ever-increasing pressure to "sell off" portions of the available spectrum, particularly in bands below 6 GHz that commercial wireless systems employ (e.g., L-band). The President's wireless broadband initiative to promote economic growth by making an additional 500 MHz available for wireless broadband within 10 years will result in RPA having to move out of sold-off spectrum. This directly affects Group 1 UAS. SUAS operations are increasingly challenged by the reassignment of RF spectrum (see Figure 19) both in the CONUS and around the world. It is difficult to coordinate approval for data link spectrum and, in many cases, impossible to comply with host-nation requirements using the existing hardware. A major RPA spectrum proposal was introduced at the WRC-12; under that proposal, the U.S. delegates advocated for a dedicated LOS RPA C2 band in the C-band spectrum. While the final results are not known at this point, this issue has general support and would provide an LOS C2 band where RPA would have primary operating status. In addition, the U.S. delegation to WRC-12 proposed establishing RPA C2 status in the current Ku-band commercial SATCOM band used by DoD RPA; this would provide increased regulatory protection for C2 of RPA, in non-segregated airspace, if approved.

For day-to-day operations, planning is an essential function needed to help deconflict operations. Close coordination between RPA system developers (e.g., AFRL, SPOs), MAJCOMs, Air Force Spectrum Management Office (AFSMO), and the combined forces commander frequency managers is critical to successful development and employment of capability. Operators should be aware of the frequency characteristics of RPA, the bandwidth requirements for sensor products, communication relay throughput, platform emission patterns, and characteristics for all links, as they relate to the electromagnetic environment where they plan to operate. Knowledge of these factors will enable operators to clearly articulate RF requirements to the frequency manager for frequency allocation and deconfliction.

As an example of current challenges, many of the RPA systems in use today were developed with legacy data link equipment that offers reduced costs and shorter development periods but is not primarily intended for air-to-ground aeronautical mobile applications. Hence, many DoD RPA must operate on a low-priority non-interference basis, which can result in having to terminate operations altogether, potentially with little to no notice. To mitigate this, new programs need to plan ahead for spectrum supportability of their primary and alternate data link communication solutions.

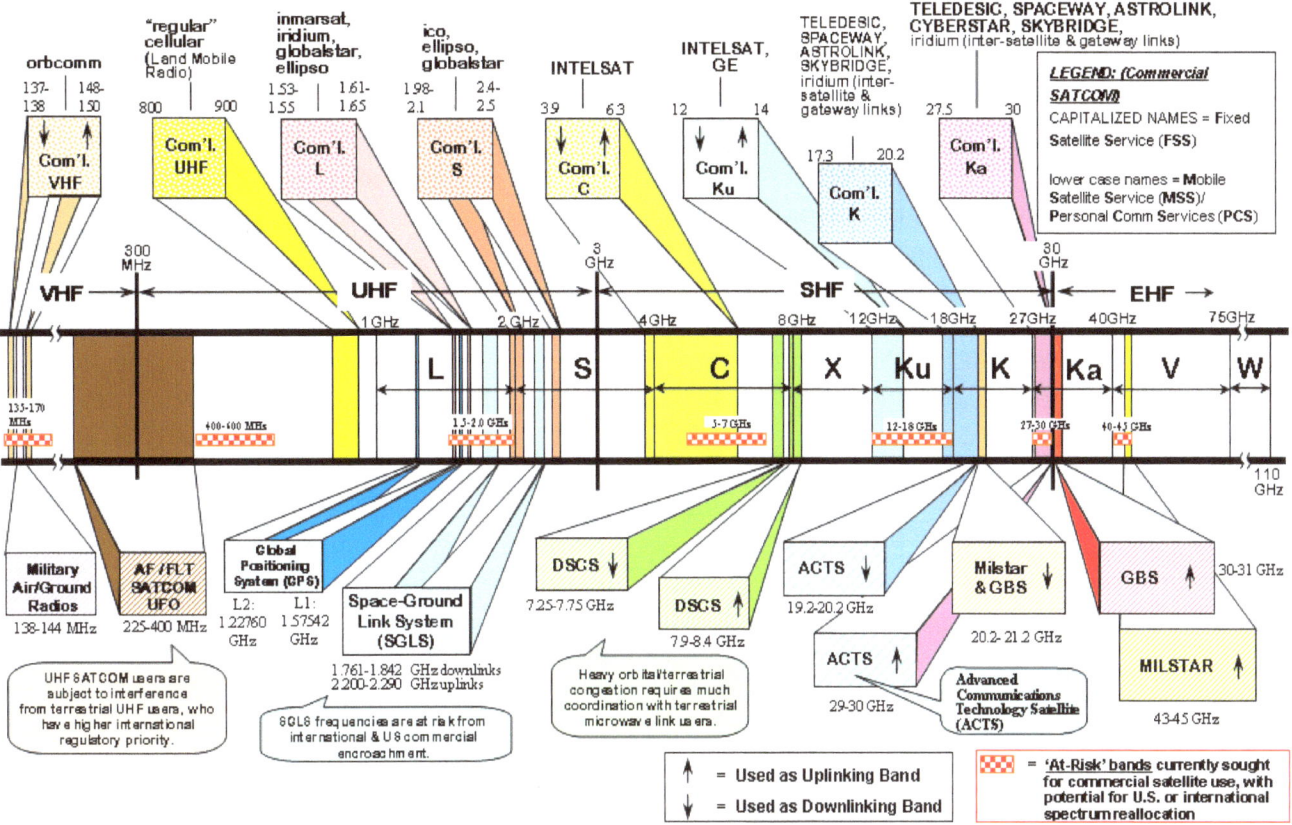

Figure 19: Radio Frequency Spectrum

RPA operators who use LOS links for control of RPA and receipt of sensor products must also coordinate with the appropriate spectrum manager to deconflict with other users. Planners must consider emitters in the local areas of both the GCS and aircraft to avoid mutual interference with other systems. For BLOS operations, regulatory requirements, potential interference and availability of military or commercial satellite access and coverage should be considered. Operators must have a solid understanding of the spectrum environment and bandwidth limitations to maximize effective use of all assets. As outlined in the Air Force Aerial Layer Networking Flight Plan, AFSPC will define mission requirements for spectrally efficient networking waveforms as well as dynamic spectrum access and will implement improved spectrum efficiencies to existing waveforms to remedy this problem in the future. In the long-term, the Air Force must design RPA with spectrum flexibility through software reprogrammable radios and associated data links and through RF hardware (e.g., antenna) to provide spectrum agility.

4.1.1.6   *Dynamic Spectrally Agile and Efficient Radios and Modems*

Continued RDT&E of emerging capability for dynamic, spectrally agile radios to adapt to decreasing available spectrum is critical. DARPA's NextGen project and its follow-on Wireless Network after Next program have demonstrated the feasibility of dynamic spectrum access (DSA). DSA offers the ability to change frequency band use based on other adjacent radio actual use and nonuse of certain bands. The Joint Tactical Radio System program is investigating the feasibility of integrating DSA technologies into its system.

In addition, long-term design for RPA C2 links should include the ability to selectively segregate the C2 link from the sensor payload link to allow for movement into more spectrally supportable frequency bands. This capability will provide more flexibility and opportunities for DoD to seek improved spectrum locations, such as CONUS operation in non-segregated airspace.

4.1.1.7   *SUAS Spectrum Supportability*

In identifying a strategy for SUAS spectrum management, several factors come in to play to make spectrum a significant challenge for SUAS data links. First, as previously mentioned, the Federal Communications Commission has pledged to auction off 500 MHz of federally owned bandwidth for commercial wireless broadband use over the next 10 years. It is expected that much of that spectrum will come from lower bands. Based on their SWaP, current SUAS typically can only support data link operations in lower bands. As an example, a popularly used SUAS band, L-band, has now become unsupportable, and existing SUAS in that band are being asked to identify migration strategies to other bands. Given mobile broadband interest in these lower bands, SUAS planners can only expect this to become more challenging. SUAS developers and program offices must coordinate closely in advance with the spectrum community (e.g., SPO, MAJCOM, AFSMO, GCCs) to identify the optimal spectrum location for their systems. In addition, technological capabilities, such as software-defined radios capable of tuning appropriate bands across the limits of the available antenna and using as narrow a bandwidth as possible (e.g., < 5 MHz) with the lowest power, are required. This will provide for greater spectrum flexibility and supportability when planning for operations.

4.1.2   Bandwidth Efficiencies

4.1.2.1   *Bandwidth Management*

RPA systems of the future should incorporate the latest improvements in bandwidth efficiency. That includes following new efficient modem standards and initiatives in improved compression algorithms and modulation schemes (e.g., turbo code). Beyond technical compression of all the collected data, there are logical advances that could reduce the amount of information that needs to be sent. One example would be to incorporate logic that frequently updates information and metadata about a target's position with a more recent update but less frequently retransmits the less important or relatively unchanging background information.

The Air Force should support advancements in modem design and software or hardware that provide adaptive capabilities to optimize bandwidth use. Near-real-time requirements for large volumes of bandwidth put great stress on supporting bandwidth systems, and the desired timeliness of sensor products has a major impact on the bits per second demand across the enterprise. It is essential that sensor product timelines be fully validated by receiving customers to ensure the demand is valid and not purely a function of the sensor capabilities.

While the application of turbo codes should give some bandwidth demand relief, history has shown that demand will grow to fill the available bandwidth; at some point, either more bandwidth is required or it needs to be used more efficiently.

Erasure codes (e.g., fountain codes) also need to be explored as part of the bandwidth management and assured communication problem. Whereas improved coding, such as turbo codes, are used to deal with data corruption, erasure codes deal with an entirely different problem: missing rather than corrupted bits due to loss-of-link and transmission receive problems. Erasure codes provide redundancy in a way that the original message can be reconstructed from a subset of symbols.

4.1.2.2　Bandwidth Efficient Modems

Air Force RPA SATCOM requirements continue to grow due to both increased quantity of platforms and CAPs fielded as well as increased sensor capability. The total MQ-1/9 bandwidth increase anticipated between FY10 and FY17 is approximately 275 percent. This amount multiplied by the rough order cost per MHz would result in more than $75 million lease costs each year. To offset some of that cost, the Air Force is considering fielding WGS MILSATCOM Ka-band capability with the MQ-9; however, the Air Force must also look to any and all efficiencies that will pay large dividends, such as BE modems. Currently MQ-1 and MQ-9 are using legacy modems that are effective but not efficient. Much more efficient modems and coding (e.g., turbo codes, fountain codes) are available that would provide for significant efficiency (greater than 25 percent) over current modems. Those efficiencies will save the Air Force money in commercial SATCOM leases or reduce the demand, resulting in improved bandwidth availability. Considerable RDT&E funding is required to support coding changes and new field programmable gate array chips in SATCOM modems. This requires an initial investment by individual programs but would likely lead to overall cost savings to the Air Force and DoD in the long term.

4.1.2.3　Improved Compression Technology

As sensor performance and associated bandwidth use continue to increase, the Air Force should also pursue technologies that will reduce the overall bandwidth demand. RDT&E must continue to improve compression of sensor bandwidth with acceptable levels of sensor product quality. Platforms must work closely together to leverage sensor compression improvements that could be applied across multiple platforms.

4.1.2.4　Onboard Storage and Processing

In conjunction with improvements in sensor compression technology, streamlined onboard storage and processing must also be considered. Improved methods of storing, processing and disseminating onboard data can offer reductions in overall bandwidth demand on the communications enterprise. Program offices, MAJCOMs, and the requirements process must also pay close attention to timing requirements for delivery of products. Most data is currently tagged for real time or near real time, which puts maximum demand on available bandwidth, increases cost and reduces available spectrum. Where mission requirements allow, storage or buffering of data and more flexible time constraints can significantly ease the bandwidth demand and should be considered. In addition, consideration should be given to pursuing technologies that selectively transmit small portions of relevant data rather than the entire data set to reduce required bandwidth (e.g., chip-out or transmitting only changed pixels).

4.1.2.5　Standard and Bandwidth-Efficient LOS Data Links

Air Force RPA have historically been fielded directly from Advanced Concept Technology Demonstrations using the analog data link technology available during initial development. While the analog links have been effective in providing needed capability, they are considered legacy and not interoperable across the joint warfighting environment. As major programmatic changes are made, standardized data link protocols (e.g., standard common data link [STD-CDL]) must be implemented to better support interoperability and establish a baseline among user subsystems, such as RVT and across Air Force and other service platforms. In addition, the Air Force must look ahead to data link improvements, such as bandwidth-efficient common data links (BE CDL) that significantly improve the use of available spectrum when compared to STD-CDL. While backwards compatibility with legacy CDL standards is a consideration, BE CDL is already being implemented by other services (most notably the Army).

4.1.3   Assured Communications

4.1.3.1   Encryption

An essential element of protecting the integrity of our communications links is accomplished through encryption. While it does not mitigate other threats, such as jamming, encryption can mitigate interception and manipulation of both C2 and sensor data links. Recently published DoD Instruction (DoDI S-4660.04) specifically identifies the requirements for RPA encryption, including:

- Encrypting aircraft control data links of RPA that carry kinetic weapons with National Security Agency (NSA) Type 1.
- Use of KGV-135a with common data link waveforms for encryption supporting up to Top Secret communications, and use of Advanced Encryption Standard (AES) for unclassified communications.

> *"I direct the services to begin work immediately to develop and implement technical solutions, as defined by the UAS Encryption Task Force in conjunction with the ISR Task Force. These technical solutions are required to protect the exposed signals on all manned and unmanned ISR aircraft and associated ground support equipment, such as remote video terminals and ground control stations. Where possible, the service shall direct the efforts to support the longer-term migration to the full digital and NSA-approved Type 1 encryption architecture."*
>
> —*John J. Young, Jr*
> **Under Secretary of Defense for Acquisition, Technology, and Logistics**

Encryption capability requires the use of key materiel (KEYMAT) to unencrypt the encrypted signal. Currently, the delivery of KEYMAT is a relatively cumbersome, manual process, requiring separate equipment for NSA Type 1 and AES keys. In the mid-term, the Air Force must transition to a more streamlined process where a common key transfer device (e.g., AN/PYQ-10 Simple Key Loader) can be used for all types of keys. In addition, for the longer term, the Air Force should investigate more transparent methods of key distribution, including over-the-air keying/rekeying and concepts like "disposable" crypto, which provide high levels of encryption but do not require the same level of recovery response if a system is lost or compromised.

4.1.3.2   Lifeline BLOS C2

While the RQ-4 has numerous options for C2, ensuring no BLOS C2 single points of failure, MQ-1 and MQ-9 fleets have limited options for BLOS C2. To increase redundancy in C2 options for RPA, tradeoffs need to be made on cost, SWaP, and risk; a secondary "lifeline" backup C2 SATCOM option should be explored, including use of INMARSAT or Mobile User Objective System ultra-high frequency (UHF) MILSATCOM constellations as is currently being implemented by other agencies.

In the longer term, the Air Force must consider expanding the C2 architecture options beyond current point to point, including MUM teaming and use of airborne networks, wherein a manned aircraft may be able to provide C2 of an RPA, given appropriate authority, communications, and level of security. MUM refers to the relationship established between manned and unmanned systems executing a common mission as an integrated team. Provided the necessary credentials and connectivity, another node could assume C2 of an RPA to enable successful recovery or execution of other missions as needed and coordinated.

4.1.3.3   *Precision Navigation and Timing in GPS-Denied Environments*

Many of the features of assured communications rely on the aircraft knowing where it is in space and time, and most mission sets require it. The broadcast signal from space can be jammed, intentionally or not, impacting operations. Onboard RPA navigation systems should incorporate redundant PNT capabilities, such as sensor-aided navigation in the event GPS signals are lost.

The Air Force must continue to integrate remotely piloted and manned capability so that, lacking data link assuredness or autonomous operation, the Air Force will still have the ability to hold strategic targets at risk. This should include the synchronization of MUM assets and modular aircraft that have optionally piloted capability.

RPA will need GPS independence or augmentation to enable operations in contested or denied environments, including supplementing GPS with use of onboard terrain-aided navigation and enabling RPA to "fight through" GPS jamming with little to no effect on operations.

In addition, chip-scale atomic clocks and inertial measurement units based on cold atom principles or other technologies can provide low-drift PNT in the event of GPS loss. These approaches maintain GPS-like position and minimize timing uncertainties over relatively long periods after GPS signal loss. By communicating with systems outside the contested or denied environment, an RPA can receive intermittent reference position and timing information for limited updates to correct drift during long-duration denial. Technologies can enable miniaturization of such systems and support system-level network functions to negate the asymmetric advantage that GPS jamming could otherwise provide the adversary.

4.1.3.4   *Improved Data Link and Modem Technologies*

Continued RDT&E of emerging, scalable optical communications solutions are essential to support extremely wideband RPA sensor needs and provide for improved LPI/LPD characteristics compared to traditional RF links.

Transmission modes other than RF, such as laser communications, are also being explored. Lasers have the benefit of being inherently LPI/LPD but may be restricted to medium- to high-altitude air-to-air or air-to-space links due to the typical problems lasers have traversing lower atmospheric conditions. "Hybrid" laser/RF terminals that alternate between laser or RF modes based on current atmospheric conditions may prove valuable in offsetting laser communication limitations but must be cost-effective and meet SWaP limitations.

The Air Force must collaborate with industry to improve the efficiency of future radios and move away from embedded RF components instead of segregating from radio digital processors, which will likely change more frequently than RF components. In addition, as the Air Force moves toward more networked operations, the internetworking layer should be moved outside of the waveform, and routers on aircraft should be segregated from individual radios to provide efficient management of routing and bandwidth needs.

4.1.4   Aerial Layer Networking

4.1.4.1   *RPA Contribution to Aerial Layer Network Operations*

The Air Force has taken a leadership role in fielding communications capability on RPA to support moving information around the area of operations via the JALN. Based on the results of the recent JALN AoA, RPA will be integral to establishing JALN capabilities, such as the HCB or medium-altitude DARE function (see Figure 20).

Figure 20: Aerial Layer Networking

The BACN on EQ-4B Block 20 provides the first significant RPA communications relay capability to support the warfighter. In addition to BACN, considerable prototyping has been used to demonstrate the RAIN architecture and capability for distributing data across the JALN, down to ground and surface users to satisfy joint information requirements. The vision for the Air Force aerial layer networking includes—

- Extending and augmenting space and surface networks to connect, reconnect, and enable the collaboration of warfighters executing specific missions and tasks in a joint operations area
- Providing the warfighter with information exchange capabilities that preserve and expand information superiority
- Integrating with, contributing to, and using the JALN
- Integration of joint net-centric components, such as knowledge management, network management, and information assurance
- Leveraging increased system modularity, open standards, and economies of scale—aerial layer networking considerations derived from a combination of operational requirements and networking objectives institutionalized in collective vice disparate capacities throughout our requirements, programming, and acquisitions processes
- DOTMLPF adaptations.

RPA can be used to enable this vision through fielding of initial aerial layer network capability, such as BLOS C2 and BACN, and evolve to deliver increased communications dissemination and C2 and ISR content to the warfighter, including scenarios where SATCOM is degraded. JALN and RPA communities must continue to work with industry to identify networking challenges (e.g., layered IP addressing) and solutions to enable capabilities, such as mobile ad hoc networking.

4.1.4.2   *Remotely Accessible Internet Protocol Network*

RAIN is an enabling architecture to support the future of Internet on the battlefield. RAIN leverages an automated software application enabling joint wireless networking using existing communication hardware infrastructures. The vision is for warfighters to be interconnected for advanced collaboration and information exchange. Currently, RAIN leverages aircraft pods specifically developed for MQ-9 as well as fighter SNIPER and LITENING tactical pods. RAIN has the potential of establishing distributed collaboration teams across functional domains that will enable intelligence providers, exploiters, and supported units the ability to communicate and transfer voice, data, and video on the battlefield in real time. This specifically brings reachback support to disadvantaged users, such as the forward TOC, convoys, and dismounts. RAIN exploits varied communication architectures to maintain flexibility.

4.1.4.3   *Battlefield Airborne Communications Node*

In GCCs today, real-time and near-real-time communications are a necessity to provide commanders the ability to C2 forces. Unfortunately, many of the communications systems currently employed in the joint and coalition battlefield are not fully interoperable and do not operate synergistically. CCDRs have identified multiple capability gaps, including LOS/BLOS bandwidth, infrastructure, and disconnected operations—all of which are problems that extend to nearly every mission area. These capability gaps have been identified in the past by several documents, two of which include the Net-Centric Operational Environment (NCOE) Joint Capabilities Document (JCD), Version 1.0 15 December 2006, and the JALN ICD, 27 August 2009. At its core, the JALN ICD documented the requirement for a HCB supporting all warfighters in a joint operating area. The capability provided by BACN fulfills enduring needs identified in both the NCOE JCD and the JALN ICD. The Air Force must ensure that the JALN architecture is interoperable across the enterprise.

4.1.4.4   *Bidirectional VORTEX*

VORTEX is a multiband capable radio and supports AES/NSA Type 1 encryption. VORTEX Phase I is a USSOCOM PoR managed by Aeronautical Systems Center (ASC), Medium-Altitude Unmanned Aircraft Systems (ASC/WII) and includes encrypted C and S Band (AES/NSA Type 1), is multiwaveform (e.g., CDL, ROVER 466ER, VORTEX Native Waveform), dual channel, 5-band (C, L, S, Ku, and UHF), and transmit only. The first retrofits for MQ-1 started in the third quarter of FY12 with additional retrofits for MQ-9 occurring in the third quarter of FY13. VORTEX Phase II is an ASC/WII PoR and includes an integrated software solution for the MQ-1 that will enable frequency and crypto control from the GCS.

The VORTEX program is a spiral development effort that integrates hardware to support duplex LOS operations and tactical IP communications using the NET-T firmware upgrade as a primary means of FMV LOS dissemination. Future upgrades include software that enables in-flight configuration changes and improved encryption technologies.

4.2   Interoperability

The Air Force vision for interoperability requires horizontal integration and commonality across remotely piloted, manned, and ground support systems. In support of the vision, many key areas must be addressed, including implementation of standards and interoperability profiles for data, data links, and interfaces (e.g., Standardization Agreement (STANAG) 4586); leveraging SOAs to share mission information; and improved C2 architectures to move away from stove-piped systems and enable collaborative employment concepts.

Where developed proprietary software is already in place, open interpreters enable the seamless integration of legacy aircraft with RPA as well as joint and coalition assets. In addition to standards and architecture development efforts, TTPs must be developed to disseminate actionable intelligence across joint communications boundaries.

> *"The absence of standardized TTPs for information exchange between UAS platforms and UAS users has limited the full potential of a critical operational capability."*
>
> *– Lt Gen Francis H. Kearney III, Deputy Commander USSOCOM*

Furthermore, actions to enable enterprise interoperability must include institutional verification and validation mechanisms (e.g., enterprise testing and comprehensive test events) to assess enterprise interoperability (and gaps) prior to fielding capability.

4.2.1   Levels of Control

Interoperability is a foundational need to share information and control. Multiple levels of control are feasible and allow the transfer or delegation of control of different functions for RPA and SUAS to other operators or customers. STANAG 4586, formally ratified by NATO in 2002, defines five levels of UAS control as shown in Table 3. Current Air Force efforts focus on the level of interoperability required to perform the mission versus specific levels of control. These levels of control may be exercised in the direction and control of UAS operations ranging from receipt of information or payload control to launch and recovery functions.

Table 3: Levels of Control

Level	Description
1	Indirect receipt/transmission of UA related payload data
2	Direct receipt of ISR/other data where direct covers reception of the UA payload data by the RVT when it has direct communication with the UA
3	Control and monitoring of the UA payload in addition to direct receipt of ISR/other data
4	Control and monitoring of the UA, less launch and recovery
5	Level 4, plus launch and recovery functions.

4.2.2   Standard Data Links and Interfaces

As the Air Force moves to better define the plan for enterprise interoperability and as DoD standards and USIPs are further developed and tested, there is a growing need to do the following:

- Implement DoD standard, interoperable data links and waveforms.
- Establish an Air Force lead for RPA common systems integration (CSI) management.
- Adopt DoD USIPs.
- Develop Air Force RPA IOPs.
- Establish and sustain an Air Force RPA CSI office to manage communications interfaces and interoperability from an enterprise perspective.

In line with other service CSI efforts, the Air Force does not currently have an equivalent CSI office with the ability and span of authority to affect interoperability efforts across the enterprise of manned, remotely piloted, and ground ISR systems. In addition, the Air Force should establish an Air Force interface control working group and interface control process to plan and orchestrate updates to interfaces affecting enterprise capability and to publish Air Force IOPs to articulate, to industry and other services, specifically how Air Force platforms will implement USIPs.

Under the auspices of the OSD Acquisition, Technology and Logistics UAS Task Force's Interoperability Integrated Process Team (IPT), USIPs were established to support joint interoperability requirements by creating specific points of "capability-based interoperability." The purpose of a USIP and its associated interoperability profiles is to define profiles of standards sufficient to guarantee levels of interoperability that enable a specific mission capability.

USIP 1.1 (LOS FMV): Developed as the first interoperability profile to tackle LOS transmissions of motion imagery (e.g., FMV) using STD-CDL. Most commonly used to transmit FMV from an RPA to an RVT, USIP 1.1 simplifies the 700+ page STD-CDL specification to a concise 30-page document that provides the guidance to achieve interoperability between compliant modems.

USIP 1.2 (BLOS FMV): Addresses RPA transmissions over SATCOM and is focused on driving similar platforms (e.g., Predator, Reaper, Gray Eagle) to common SATCOM transmission standards.

USIP 1.3 (BE LOS): Optimizes RPA LOS transmissions from current STD-CDL, which is effective but relatively inefficient in its use of bandwidth. For example, commonly used STD-CDL data rates are 2.0 Mbps and 10.71 Mbps. When there is a 3 Mbps FMV requirement, the only standard link choice is 10.71 Mbps, wasting a substantial amount of already scarce spectrum. BE CDL instead offers numerous incremental data rates (e.g., 512 Kbps, 1.0 Mbps, 2.0 Mbps, 4.0 Mbps, 8.4 Mbps, 10.0 Mbps) for a more customized allocation, freeing excess bandwidth for other users.

USIP 1.4 (Wide Area Sensors): Establishes standards for interfaces to and from a WAS, such as standard video control interfaces from a ground user (e.g., ROVER) to a WAS. In addition, the Air Force is working to develop a gigabit CDL standard that should be incorporated within USIP 1.4 as the new WAS wideband LOS data link standard.

USIP 1.5 (Weaponization): Standardizes weapons store and data link interfaces across services, based on the existing Army Weapons Interoperability Implementation Guide.

4.2.3   Sharing Mission Information

The Air Force currently has a multitude of manned and remotely piloted ISR platforms that support operational missions and continuation training flights in the CONUS. RPA crews must have access to a dynamic operational picture across the ISR enterprise leveraging the Air Force DCGS PED for active duty, ANG, and AFR ISR missions.

A UDOP must be created to fuse intelligence data across all domains. This UDOP should incorporate Geospatial Information System (GIS) and command, control, communications, computers, intelligence reconnaissance and surveillance (C4ISR) for both manned and remotely piloted assets. In addition to GIS and C4ISR, the RPA community's UDOP must be open to information and data contained in the Theater Battle Management Core System, Command Post of the Future, Blue Force Tracker, Global Command and Control System, DI2E, and other similar database systems.

Surveillance Intelligence Reconnaissance Information System (SIRIS) is a QRC currently used in the RPA community to overcome UDOP combat collaboration and intelligence fusion challenges, primarily using Google Earth and Falcon View as its map viewer (see Figure 21). SIRIS enables fusion and collaboration through persistent ISR data provided by industry standard web server scripting and database interfaces. Data brokering is handled through an application programming interface (API).

Users can access and share data across multiple secure domains, and this capability requires the synchronization of operational and intelligence data within a UDOP.

Zeus is another near-term UDOP solution used by RPA crews to provide mission participants with a complete real-time SA picture and integration into tactical data link networks. It incorporates multiple display configurations (including three-dimensional), a browser-based display, and support for third-party displays (such as Google Earth and Falcon View). Integration into military tactical data link networks such as Link-16 enables RPA to be active participants in the battle space. A real-time correlated and integrated track picture, created from multiple radar, track, and data link sources, is displayed in conjunction with current and forecasted weather, imagery, charts, FMV, and airspace information (see Figure 22). Currently, Zeus is not certified for operation in the NAS; as a result, its use in the training environment is limited. This limitation significantly impacts RPA training as Zeus is widely used for theater operations.

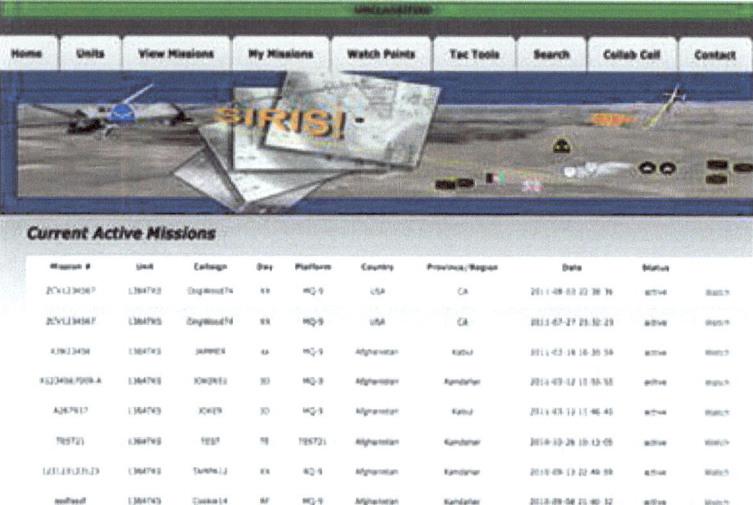

Figure 21: SIRIS Mission Display

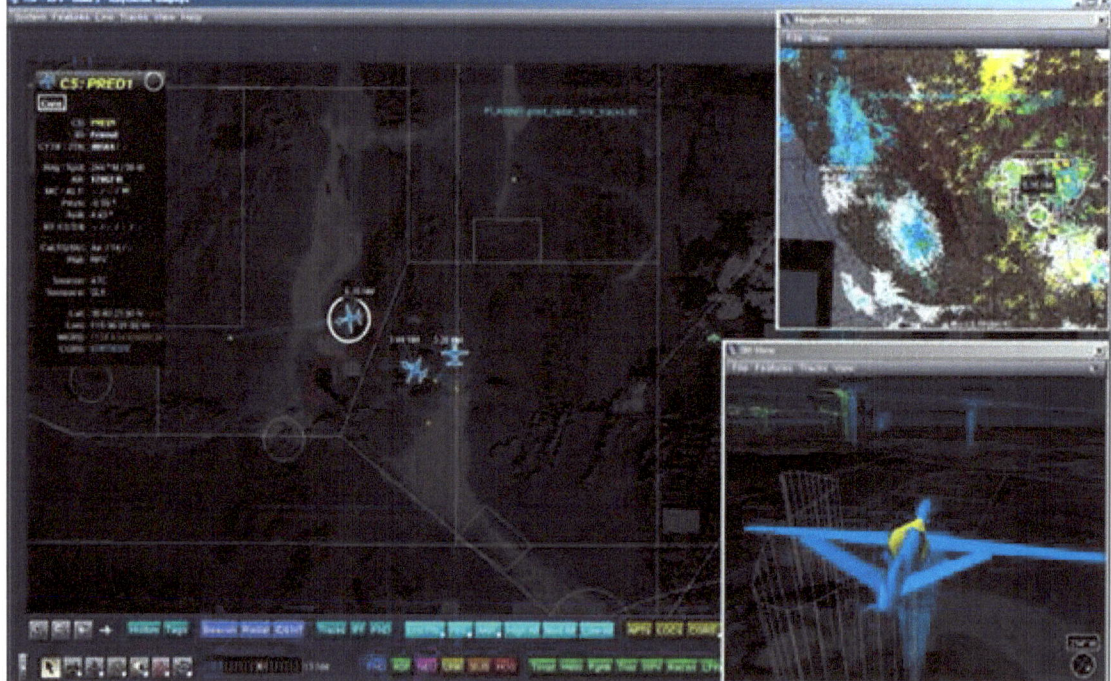

Figure 22: ZEUS Mission Display

4.2.4   C2 Architecture

The key C2 enablers to support interoperability in future operating environments include common control architectures and an alternate means for C2 of aircraft beyond the existing RSO model designed to support today's conflicts. Furthermore, improved autonomous behaviors will enable one pilot to control multiple aircraft similar to the way a flight lead in a four aircraft formation directs mission execution today in manned platforms.

4.2.4.1   *Common Control Systems and Control Architectures*

The Air Force has begun collaboration with the Navy on a joint concept for common control systems and control architectures that could serve as the model for appropriate future development of common control systems and architectures between services. This joint concept is intended to provide direction to the services and summarizes Air Force and Navy near-term, mid-term, and far-term efforts to move toward a joint solution for UAS/RPA control systems and architectures. In the future, we envision the services will work together to—

- Publish a joint, non-proprietary COTS framework, configuration descriptions, and interface definitions to efficiently allow industry partners to develop common architecture services and applications
- Establish joint governance procedures, policy, and oversight of the COTS framework and interface definitions applicable to UAS/RPA control systems and architectures for Air Force and Navy systems
- Jointly develop common internal/external interfaces and standards (e.g., STANAG, USIP, IOPs) that will promote interoperability of control systems
- Ensure that development of control system software services and applications is aligned with the common architecture strategy and leverages existing Air Force and Navy efforts (including UCS and UCI) where practical
- Build control system software services and applications with government data rights to a common architecture model with standardized interfaces
- Develop and field control systems with maximum reuse of services, applications, and common components, thereby reducing redundant development efforts and total ownership costs
- Apply common control standards and interfaces across the family of RPA
- Implement UAS operations standards.

4.2.4.2   *Alternative Communication Paths—Beyond RSO*

The current RSO model primarily uses mission C2 from CONUS; however, the potential exists to use alternative methods for control. Mission control elements could be positioned in theater, and either ground or air-based GCS LOS and BLOS links could be employed through airborne relay nodes.

Examples of airborne C2 include manned aircraft controlling SUAS, manned aircraft controlling larger (e.g., Group 4 or 5) RPA, and RPA controlling SUAS and/or lethal air-launched SUAS. The future vision includes the loyal wingman concept where one manned aircraft controls one or more RPA as a part of the mission package (e.g., AI, C2, SEAD). This platform would contain mission control elements for RPA using UCI (and potentially UCS) technologies to ensure interoperability across multiple RPA types. It is important to note that the term "controlling" does not necessarily imply flight control, but rather control of tasking the aircraft in a cooperative fashion to achieve mission goals as a part of the mission package.

Alternative "lifeline" C2 links are also possible, including low-cost, narrowband SATCOM (e.g., INMARSAT) and low-band LOS RFs. Use of shortwave (high frequency) offers extended reach and has recently been improved with Digital Radio Mondiale technology to overcome challenges inherent in this band (e.g., fade, drift, interference). While the bandwidth may not be practical for payload data, it may be sufficient for basic C2 when paired with appropriate levels of autonomy.

4.3   Autonomy

As RPA operations continue to grow in demand, the drive toward greater efficiency and effectiveness must continue. This demand presents both challenges and opportunities to operate RPA with decreased reliance on human interaction. Increased use of autonomous behaviors such as MAC may provide a means to achieve efficiencies and enable technology growth for advanced concepts such as swarming and MUM teaming.

> *"Instead of viewing autonomy as an intrinsic property of an unmanned vehicle in isolation, the design and operation of autonomous systems needs to be considered in terms of human-system collaboration."*
>
> *– Defense Science Board Report, July 2012*
> *The Role of Autonomy in DoD Systems*

Autonomy will also be applied to process onboard information and to enable some maintenance functions such as improved/increased diagnostics/prognostics to accelerate fault isolation, troubleshooting, repair, and potentially aircraft ground refueling. Swarm technology will allow multiple RPA to cooperatively operate in a variety of lethal and non-lethal missions at the command of a single pilot.

4.3.1   Multi-aircraft Control Technology Improvements

MAC applies today's technology to automate basic mission profiles with a man-in-the-loop capability to increase operational efficiency. Utilizing MAC as an employment method is mission and environment dependent. As this capability develops, it is expected to significantly increase the number of aircraft controlled based on the complexity of flight profiles and missions.

The near-term concept of swarming consists of a group of semi-autonomous aircraft monitored by single or multiple operators. Swarm technology will allow the commander to monitor the aircraft both individually and as a group. The aircraft within the swarm may fly semi-autonomously along a pre-programmed route to an area of interest while also avoiding collisions with other aircraft in the swarm. These aircraft will automatically execute mission tasks, including processing imagery requests from disadvantaged users or detecting threats and targets through the use of advanced algorithms or AI.

MUM teaming or loyal wingman technology differs from swarming in that a SUAS or RPA will accompany and work with a manned aircraft to conduct the mission. As such, RPA will play a key role in supporting manned assets. RPA could provide additional firepower, decoys, and jamming and disrupt enemy attacks through counter-UAS missions. Another example of a loyal wingman is a Group 5 RPA that acts as an airborne cargo convoy or refueling asset. This highlights a critical need for loyal wingman to closely match the performance of manned aircraft in these scenarios. Future concepts should focus on formations of as many as four aircraft operating together as a part of a mission package and potentially providing audible and/or visual inputs to the flight lead regarding position, status, and emergencies. Current governance and technology do not permit efficient employment across all mission areas.

4.3.1.1   RPA Transit Operations

Using the mission control element (MCE) for transit operations is inefficient and limits the ability to provide continuous coverage of multiple target areas. Employment of the MQ-1/9 through RSO requires more than one aircraft to be airborne to account for transit time when 24x7 coverage of a target area is required. MCE crews may spend several hours controlling each aircraft in transit depending on the distance from the launch and recovery element (LRE) to the target.

Persistent 24x7 coverage of a target requires either additional personnel and equipment or the development of a new control workstation to enable multiple aircraft control during transit operations. The development of a new control workstation that incorporates improvements in HMI and increased levels of automation will enable a single crew to monitor and control multiple aircraft simultaneously while in transit.

A dedicated control workstation for RPA transit operations will enable MCE crews to maintain 24x7 coverage of a target area, or cover additional targets, for a more efficient use of available manpower and equipment.

4.3.1.2   Cooperative Automated Multi-aircraft RPA Operations

Cooperative automated multi-aircraft RPA operations (CAMARO) is a technology suite that may enable future employment concepts including multi-aircraft control, loyal wingman, and swarm. The key technological capabilities required for CAMARO-enabled operations are autonomy, cooperation, and enhanced communications.

A key enabler for this concept is a level of autonomy that exceeds what is currently implemented and available today. Aircraft must be able to conduct self-diagnostics. When the link to a human is unavailable or intentionally disabled, the aircraft must perform the appropriate checklist procedure as a pilot or operator would.

To ensure the deconfliction with other aircraft, manned or remotely piloted, the aircraft must have a mature SAA system that allows for cooperation with other airborne systems. This technology will not only aid the aircraft in airborne deconfliction but also allow participation in the cooperative environment by transmitting its current three-dimensional location and intent information. For example, many Group 5 RPA utilize pre-planned waypoints to fly some missions. Pre-programmed information should be made available to mission planners, airspace control authorities and others needing this data. The transmission of its own status and mission progress, as well as receipt of other aircraft information, requires enhanced communications technologies.

Air superiority may not be guaranteed by the first day or even week of a major contingency operation. Future capabilities must allow operation in contested and denied-access environments. Future concepts, using advanced communication technologies are needed. Examples of enhanced communications methods include "by exception" communications, using timed data bursts to transmit and receive information to and from the aircraft. Technologies that increase communication range and enable RPA to receive and transmit via a variety of communication mediums will be critical.

The RPA must be able to notify the pilot when a self-diagnosis reveals a fault or malfunction that requires the pilot to make a "return to base" decision.

These enhanced communications will increase survivability in a communications-denied area while minimizing aircraft detection. Enhanced communications are also a key enabler for RPAs in MUM teaming when in the role of a loyal wingman or as a part of a swarm.

4.3.1.3   Multi-aircraft Manager

Multi-aircraft Manager (MAM) is a prototype SA display for multiple aircraft visualization. The system will provide SA of RPA status to operations supervisors at the squadron, group, and wing as well as to other stakeholders such as the ROCC and AOCs. It provides spatial and temporal views of the fleet as well as detailed information on the individual aircraft and assigned missions. Recent efforts include continued incorporation of multiple information sources and display functions, demonstrating temporal mission displays and the porting of applicable new display functionality to common operational picture tools. Future efforts of MAM look to provide active supervisory-level oversight of multiple RPA.

The short-term application of MAM aims to provide a unitwide or fleetwide SA tool for operations supervisors. A future potential application of this capability includes management of transit operations involving multiple RPA, which could increase RPA time over target. Another potential application of MAM envisions active management of SUAS. This would enable collaborative operations of RPA and SUAS in expanded missions sets (one example is the insertion and release of multiple SUAS by an RPA to search and identify targets collaboratively over an expansive area).

4.3.1.4   Foxhunt

AFRL recently completed research in these cooperative operations as part of the Foxhunt program, where smaller AL-SUAS would work cooperatively in teams to achieve common objectives. This project aimed to develop and apply algorithms supporting cooperative offboard RPA sensing from a control station onboard an airborne "mother ship."

The Foxhunt program pursued multiple capabilities applicable to larger ground-launched RPA. In Foxhunt, RPA team to seek out targets for more intense monitoring, anticipate target movement to position other "team members" (RPA) for optimal coverage and share or hand off tasks when requested. This cooperative approach can provide multiple perspectives on the same target through coordinated point surveillance. The initial Foxhunt concept demonstrated offboard RPA teams supporting onboard sensor operators to detect, monitor and cue a "mother ship" to meet mission objectives. Though some progress was made with initial capabilities, Foxhunt has been discontinued to pursue other objectives in 2013.

4.3.1.5   Flexible Levels of Execution—Interface Technologies

Flexible Levels of Execution—Interface Technologies (FLEX-IT) is an R&D project to evaluate methods for using flexible levels of RPA automation. It will develop a framework for hosting and managing libraries of automated behaviors that RPA crews can engage individually or in orchestrated groups to form higher-level automated tasks. A major focus is to develop a means of naturally integrating human awareness, supervision and control of the automated behaviors as well as the intuitive seamless transitions between levels of automation. The project will develop a set of single ship behaviors that could be transitioned to a program office for inclusion in control system software. The project will also, as resources permit, investigate multiship coordination of automated behaviors as needed with the MAM project.

The goal was to deliver system demonstration software illustrating the framework for hosting and human interface technologies. FLEX-IT capabilities could enable future capabilities in denied airspace, collaborative MAC, and expanded RPA mission sets.

4.4   Airspace Integration

DoD is working near-, mid- and long-term strategies to achieve immediate, conditional gains in NAS access, while working toward viable long-term solutions. Priority is given to initiatives that reduce

COA requirements and streamline the FAA COA approval process. The Air Force must continue working closely with the FAA UAS Integration Office and the FAA Technology Center to ensure that progress is made to allow safe, efficient and effective NAS access.

For any military aircraft, manned or remotely piloted, to fly routinely in domestic, international, and foreign national airspace, three foundational requirements must be met. These three requirements are essential and form the foundation for UAS airspace integration. First, Title 10 of the United States Code is the legal underpinning for the roles, missions, and organization of DoD and provides authority for the military departments to organize, train and equip U.S. forces to fulfill the core duties for national defense. Airspace integration must comply with all tenets of Title 10. Second, consistent with both this statutory authority and longstanding practice, and reinforced by interagency agreements, DoD is responsible for establishing airworthiness and pilot training/qualification requirements for the military while ensuring rigorous military standards are satisfied. The third and most complex requirement, regulatory compliance, encompasses both internal military department regulations and external FAA and International Civil Aviation Organization (ICAO) flight regulations.

The Air Force must continue to develop the safest, most capable fleet possible while striving for maximum compliance with existing regulatory guidance and informing regulatory processes when changes are needed. DoD will fully leverage statutory authorities to design, test, and ultimately certify its UAS in compliance with applicable standards, regulations and orders. Development of SAA systems is crucial to expanded access to the NAS and international airspace. Air Force RPA are a critical component of military operations; and to operate routinely in domestic and international airspace, the Air Force must continue to define and develop certification standards for RPA and SUAS, ensure qualified pilots/operators continue to be trained to operate in the appropriate class(es) of airspace, and develop systems that comply with applicable regulatory guidance.

4.4.1 Airworthiness

Airworthiness is a basic requirement for any aircraft system, manned or remotely piloted, to enter the NAS. The primary guidance for DoD airworthiness certification is found in MIL-HDBK-516B, Airworthiness Certification Criteria. This document defines airworthiness as "the ability of an aircraft system/vehicle to safely attain, sustain and terminate flight in accordance with an approved usage and limitation." Airworthiness certification ensures that DoD aircraft systems are designed, manufactured, and maintained to enable safe flight. Certification criteria, standards, and methods of compliance establish a minimum set of design and performance requirements for safely flying a given category and class of aircraft. The DoD is expanding current military airworthiness guidance to include criteria that addresses component and system attributes unique to RPA. RPA-unique standards derived from NATO STANAGs (e.g., 4671, 4705 and 4703) will be reviewed and incorporated as appropriate.

4.4.2 Pilot/Operator Qualification

The DoD determines where and how it will operate its aircraft, and each service creates the qualification training programs necessary to safely accomplish the missions of that aircraft or weapon system. The standards to train and qualify pilots/operators of UAS will remain under the authority of the Service and appropriate CCDRs. While RPA pilots and sensor operators face challenges operating aircraft from a remote shelter (e.g., C2 link latency), the majority of aircrew skill sets required (e.g., communication, multitasking, airmanship) are no different from those required for manned aircraft. Therefore, the services and GCCs must apply the existing minimum training standards outlined in Chairman of Joint Chiefs of Staff Instruction (CJCSI) 3255.1 to their respective training programs to ensure the requisite knowledge, skills and abilities are addressed appropriately.

4.4.3   Regulatory Compliance

> *"Unmanned aircraft systems (UAS) are a new component of the aviation system, one which ICAO, States and the aerospace industry are working to understand, define and ultimately integrate. These UAS are based on cutting edge developments in aerospace technologies, offering advancements which may open new and improved civil/commercial applications as well as improvements to the safety and efficiency of all civil aviation."*
>
> *– ICAO UAS Circular 328 2011*

DoD has a robust process for establishing manned aircraft flight standards and procedures. However, the current ambiguity and lack of definition in national and international regulatory guidelines and standards for RPA make it difficult to know, with consistency or certainty, whether RPA can comply. In fact, some current RPA may already be operating at appropriate levels of safety; however, until the necessary RPA-specific standards, regulations and agreed-upon compliance methodologies are defined, establishing regulatory compliance for more routine operations is difficult. In the meantime, RPA operations within the NAS are treated as exceptions through the COA process.

While many operational requirements can be met using manned aircraft, many missions are more efficiently and safely accomplished using remotely piloted platforms. Technology advancements should help resolve regulatory compliance issues for UA (particularly Title 14 of the U.S. Code of Federal Regulations 91.113 containing the see-and-avoid provision); however, the level and complexity of technology required to resolve today's regulatory compliance issues may negatively affect system affordability.

4.4.4   Mid-Term NAS Access

The DoD Airspace Integration CONOPS provides a framework for common UAS practices, procedures, and flight standards in NAS and international airspace. It is intended to standardize UAS access methodologies and procedures, implement appropriate methods for compliance with see-and-avoid requirements, and inform development of an UAS Airspace Integration ICD. It established a standard suite of lost-link, lost-communications, and lost-SAA procedures for DoD UAS in all phases of flight. These procedures will help define methods for notification and the appropriate action to either regain link or recover/divert the aircraft.

The CONOPS also provides the operational and procedural construct to employ the access profiles at bases across the United States and to inform the process of basing UAS in locations OCONUS. The CONOPS builds upon the six profiles outlined in the DoD's Airspace Integration Plan, which may be used individually to access specific local airspace or integrated together to satisfy additional airspace requirements. The profiles represent a significant step forward in organizing airspace access requirements in a standard and measurable methodology.

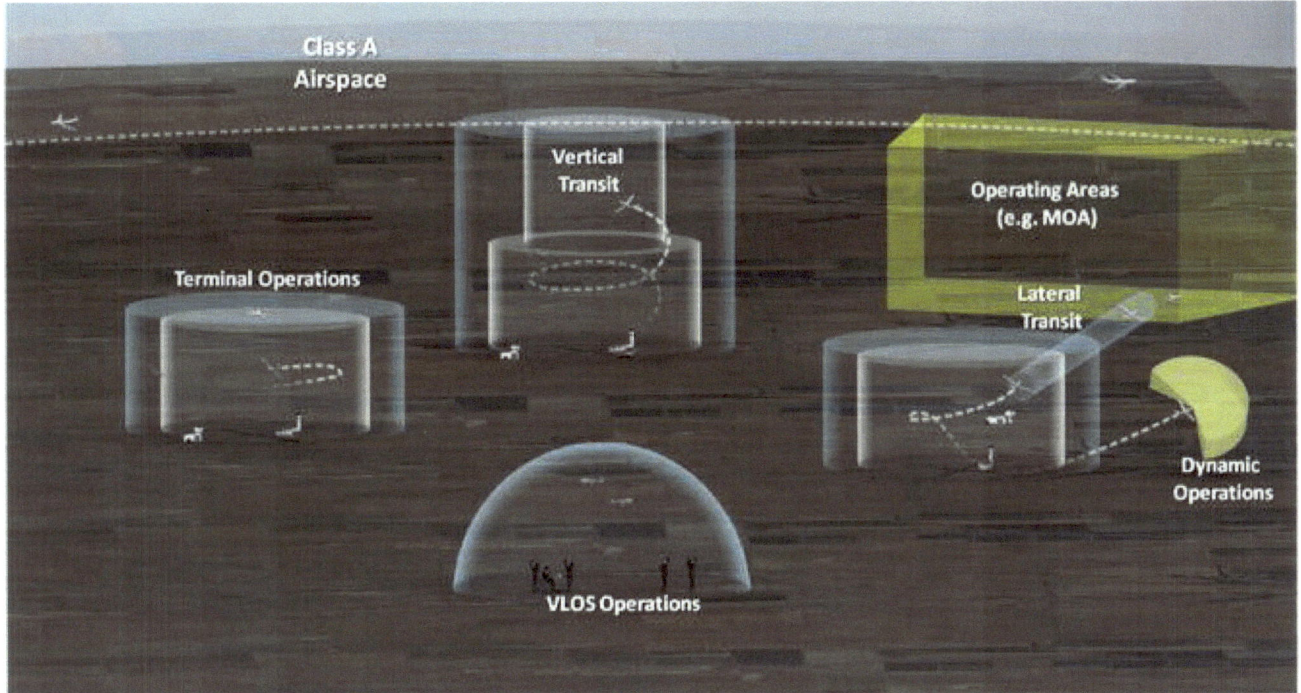

Figure 23: Access Profiles

4.4.4.1   *Ground-Based Sense and Avoid*

Mid-term NAS access can include new FAA rules regarding Class A airspace that permit RPA to operate with other aircraft. New rules may also be augmented with GBSAA technologies for access into other classes of airspace. GBSAA is a ground-based means of detecting airborne traffic and providing the necessary SA to the SUAS-O or RPA pilot. A GBSAA system includes sensors, correlation, fusion, communications, networks, logic, procedures, and user interfaces. GBSAA is a technological solution using ground sensors to detect aircraft and alert the RPA pilot or SUAS-O with suitable lead time to allow appropriate avoidance maneuvering, enabling safe and robust mission operations without risk of airborne collisions. The first DoD GBSAA for RPA went operational at the Army's El Mirage, California, facility on 26 April 2011. This system at El Mirage, California, had limited operational application and has since been suspended and moved to Dugway Proving Grounds due to continuing regulatory differences. The Air Force must continue to build on the initial framework to increase access through a more robust detection and alerting system.

Mid-term GBSAA efforts are focused on developing methods to provide aircraft separation within a prescribed volume of airspace using a ground-based system that includes sensors, displays, communications, controls, and software. The complete system will include HMI displays, controls, and software algorithms necessary to provide recommended or perhaps automated maneuvers. GBSAA solutions will incrementally relieve restrictions on existing COAs and facilitate RPA/SUAS training in the NAS. This effort is establishing requirements, gathering data, performing modeling and simulation, testing and verifying collected data, and obtaining airworthiness approvals, as appropriate. GBSAA can particularly benefit SUAS where SWaP limitations prevent larger more complex ABSAA solutions from being easily incorporated. Such a system should be a low-cost, scalable, deployable system capable of responding to Title 10 and Title 32 National Guard requirements for DSCA.

4.4.5   Long-Term NAS Access

Current UAS are built to different specifications for different purposes; therefore, showing individually that each system is safe for flight in the NAS can be complicated, time consuming and costly. Routine access will happen when DoD establishes and certifies an acceptable level of safety for UAS and manufacturing standards meet that threshold. Once developed, standards for UAS acquisition will improve system interoperability and yield cost savings.

> *"The safe integration of UAS into non-segregated airspace will be a long-term activity with many stakeholders adding their expertise on such diverse topics as licensing and medical qualification of UAS crew, technologies for detect and avoid systems, frequency spectrum (including its protection from unintentional or unlawful interference), separation standards from other aircraft, and development of a robust regulatory framework."*
>
> *– ICAO UAS Circular 328 2011*

One key long-term objective is the fielding of a certified ABSAA solution. Airspace access will be similar to manned aircraft; though certification requirements have not been established, the current level of technology and automation demonstrated may be sufficient.

4.4.5.1   *Airborne Sense and Avoid*

ABSAA development efforts are focusing on onboard capability to perform both self-separation and collision avoidance to ensure an appropriate level of safety. Current programs have phased validation schedules for due regard, en-route/Class A, and divert/Class E/G operations as technology innovation and integration allow.

The ABSAA system must provide autonomous maneuvering to maintain self-separation/collision avoidance in the event of lost link and provide for pilot-in-the-loop capability as operations dictate. The Air Force and Navy should leverage a common functional baseline for the RQ-4 Global Hawk and MQ-4 Broad Area Maritime Surveillance (BAMS) aircraft as initial demonstration platforms for ABSAA. The capability should be extended to other RPA and SUAS where it provides strategic benefit and can be seamlessly integrated with GBSAA technology as part of a SoS solution.

Autonomy will mitigate loss of data links and will ensure avoidance maneuvers are executed in time to avoid breaching defined collision thresholds. The system will be designed to be modular, open architecture, non-proprietary (Government data rights), and multiaircraft capable, providing commonality and interoperability across RPA. These technologies will further be scalable to SUAS. The heart of the Common-ABSAA capability will be a sensor/platform-agnostic data integration/fusion engine and its accompanying collision avoidance algorithm. The Common-ABSAA will use sensor and aircraft personality modules to control the interfaces for multiple aircraft sensor and MDS configurations. Onboard non-cooperative sensors such as radar, EO/IR, LIDAR, and acoustic, as well as cooperative sensors such as Traffic Collision Avoidance System (TCAS) and ADS-B, can be mixed and matched to meet current and emerging requirements in all classes of airspace. ABSAA will pursue commonality with GBSAA solutions to minimize duplication; leverage lessons learned; maximize common training, maintenance and documentation; and provide a seamless ABSAA and GBSAA HMI for the pilot/operator. Greater ABSAA/GBSAA commonality will reduce life cycle costs, build efficiencies, and promote ease of future upgrades.

4.4.5.2   *Autonomous Terminal Area and Ground Operations*

The objective of an autonomous terminal area and ground operations (ATAO) effort is to develop a control architecture that enables RPA to operate in the terminal area with specific emphasis on ground operations. An ATAO-compliant RPA will receive, read back, and perform all commands from ATC in a timely manner. In the case of unclear instructions or if the RPA determines ATC has given an incorrect command, the RPA will obtain clarification. After receiving commands from ATC, the RPA will complete the assigned task in a human-like fashion, using pre-programmed autonomous behaviors. Similar to human operators, the RPA will combine observed information with holistic modeling of typical airfield operations to predict the actions of other aircraft. The RPA can stop and query the controller if it detects an unsafe situation developing.

4.4.6 Sensors and Payloads

To meet near-term warfighter mission requirements and support the core functions identified in this vision, new sensors, payloads, and weapons will be required. Some examples of ongoing developmental efforts are found in the following sections:

4.4.7 Enhancements to Gorgon Stare

Gorgon Stare provides day/night continuous broad-area motion imagery to find and fix targets within the field of view. The capability combines real-time SA for SCAR or cross-cueing other sensors, with persistent video recording for forensic analysis and pattern-of-life study. Algorithms that would detect moving targets will be integrated via standalone systems at the appropriate ground ingest point.

4.4.8 Dismount Detection Radar

The dismount detection radar (DDR) fills a gap in radar collection/tracking of dismounted personnel. The DDR system will have the ability to simultaneously transmit and disseminate DDR sensor data to real-time users, intelligence exploitation nodes, and DoD data repositories. The DDR system will support Level 3 control, allowing control and monitoring of the UA payload by a designated tasking authority, and will augment MQ-9 Block 5 Reaper's hunter/killer mission. MQ-9s equipped with DDR will provide persistent detection and tracking of vehicles and dismounts in day/night, all-weather conditions and will be integrated into DCGS. The system will be BLOS and LOS capable and will provide commanders and intelligence analysts with reconnaissance to support SA, target identification, time sensitive targets, target engagement, BDA, dismount patterns of life, and intelligence preparation of the battlespace. In a Title 32 domestic response, this system will provide an incident awareness and assessment role for authorities during natural disasters. There are also potential applications for border security operations. DDR capability is planned for operations by the end of FY14.

4.4.9 Airborne Cueing and Exploitation System Hyperspectral

The Airborne Cueing and Exploitation System Hyperspectral (ACES HY) is an advanced infrared sensor that provides the capability to find hard-to-detect targets in a wide search area by collecting and processing target data from the visible (.4 mm) to shortwave (2.5 mm) spectrum. The first production sensors were delivered in March of FY12. ACES HY sensor and processor design will be integrated into the configuration baseline and PED upgrades planned to enable data exploitation.

4.4.10 High-Definition Digital Video Architecture

Modifications to Block 15, 30, and 50 MQ-9s GCS will incorporate high definition (HD). The full-scale HD Digital Video Architecture will be incorporated into Block 5 MQ-9s. AFSOC modified four aircraft and three GCSs to perform HD operations in December 2011 per JROCM 066-10. In April 2012, AFSOC modified three additional aircraft and one GCS to support HD operations. The first phase implements target location accuracy (TLA), which will support enhanced data exploitation tools, including real-time display of target coordinates, digital data archiving, digital video recorder playback capability, and image mosaicking. The first phase also integrates the Raytheon Community Sensor Model (CSM), which separates "key length value" metadata for LOS/BLOS links, integrates CSM in the GCS, and provides "near-frame synchronous" metadata. The final fielded TLA (-3) MTS-B turret will enable "frame synchronous" metadata and 720p HD IR.

4.4.11 IP Infrastructure/Generic Payload

The IP Infrastructure program implements IP/Ethernet for RPA payloads and supports flight systems/payload separation, which is critical to airworthiness. In addition, the Generic Payload over IP implements command architecture to allow for plug-and-play payloads on the MQ-9 and third-party API.

4.5   Training and Improved Human Interfaces

To the maximum extent possible, common HMIs should be fielded. Adopting and maintaining standardized displays and operating systems reduces developmental and training costs across the RPA fleets. Common look and feel interfaces could be shared by a variety of RPA operators performing like missions or employing the same group of aircraft.

The focus of the Air Force has been to rapidly field capabilities to meet mission needs. As we transition to steady-state CONOPS, we need to invest in improving the training and ground control system interfaces to increase the effectiveness of our pilots and mission payload operators. It is imperative that DOTMLPF actions are identified to adequately train and qualify RPA crews for future missions beyond armed ISR such as CAS, SCAR, and PR. In addition, a shift to more traditional routine training and exercise participation is required to increase mission readiness and allow for formalized training syllabi and an evaluation methodology.

4.5.1   High-Fidelity Simulation

The Air Force's MQ-1 and MQ-9 Predator Mission Aircrew Training System (PMATS) is a high-fidelity simulation system that models Predator and Reaper aircraft, sensors, and weapons for initial and mission qualification training. The Air Force plans to procure 48 additional PMATS from FY12 to FY14, to upgrade legacy PMATS to the current block, and to ensure procurement meets the demands of the expanding enterprise. In addition, the Air Force continuously improves PMATS with software updates to keep it aligned with aircraft and sensor fidelity requirements. Across all Air Force platforms, about 40 percent of aircrew qualification training is in a simulated environment. The Air Force uses "distributed mission operations" to link its high-fidelity simulators, allowing advanced training activities between geographically separated units and enabling a more constructive training environment. This distributed mission network broadens the training environment, enabling improved operational mission rehearsal, CONOPS development, and combat readiness with minimal impact to combat operations. MQ-1 and MQ-9 are not scheduled to connect into the distributed network until at least FY14/15. The RQ-4 Powertrain Systems Analysis Toolkit (PSAT) is a part task trainer that simulates basic operation and emergency procedures for Global Hawk pilots during initial qualification training. Because the PSAT is not a high-fidelity or networked simulator, mission qualification training can be completed only during real-world missions. While this offers excellent training for specific mission areas, it does not represent the full spectrum of operations that crews must train to, nor does it incorporate integration with other platforms in a major combat engagement. A high-fidelity, networked simulator is currently not programmed for RQ-4.

4.5.2   Dynamic Maneuvering

The Dynamic Maneuvering (DYNAMAN) project envisions a mature capability for predictive flying of RPA while in RSO to overcome the communications link latency by artificially presenting the forecast aircraft attitude to the pilot based on control input.   Initially aimed at MQ-1, MQ-9, and MQ-X, DYNAMAN seeks to improve mission performance and increase options in CAS, IW and tactical ISR environments through providing the ability to dynamically fly the aircraft.   Basic capabilities were demonstrated in the AFRL Human Performance Wing's FY10 DYNAMAN II research project.   Recent efforts have refined the predictive algorithms and heads-up display symbology.   They have also incorporated an MQ-9 flight model and evaluated the systems resiliency to intermittent links.

Future applications of DYNAMAN allow rapid and reliable pilot control of aircraft for unforeseen situations, improved or new tactics, and expanded mission sets.

4.5.3   Dynamic Immersive Mission Environment

The Dynamic Immersive Mission Environment (DIME) project focuses on the development of a virtual environment in the GCS to significantly enhance aircrew SA and the ability to react precisely.   DIME aims to leverage the unique distributive, persistent, and interoperable attributes of remotely piloted aviation to accelerate and significantly enhance mission execution with RPA.

Since the GCS can link to nearly all information sources, massive amounts of information flow to the crews of large RPA for ingestion by each crew member to create the operational environment.   DIME seeks to present that information in the surrounding virtual environment for the crews in a visually natural way.   These visuals will include high-fidelity depictions of terrain, friendly and enemy locations, weather, surface-to-air threat domes, airspace blocks and airborne traffic—all in a single three-dimensional picture surrounding the pilot.   This promises to decrease training time and cost, increase crew proficiency, greatly diminish information saturation, and heighten aircrew awareness to levels potentially far beyond legacy manned aircrew standards.   As a result, with mission-focused RPA, DIME could enable an expanded RPA mission set leveraging its unique capabilities for remotely piloted aviation to more effectively accomplish the Air Force's core mission sets.

4.6   SUAS Capability Development

4.6.1   Small UAS Technology

The Group 1 capabilities shown in Figure 24 highlight the current SUAS efforts within AFSOC and the Air Force; those outlined in black are funded efforts, red are unfunded, and yellow are technology demonstrations.   Currently, the WASP is the only Air Force PoR with a total on hand of 238 Block III and 20 Block IV systems as seen in Figure 24.   Shown in the second row, the Air Force was the first to acquire and field a ScanEagle system.   However, without baseline funding as a PoR, Air Force Security Forces will not continue ScanEagle operations indefinitely, even though the system proved its combat effectiveness in theater.   Currently, Air Force ScanEagle operations are sustained using what remains of a $27 million OCO funding line, supporting Air and Space Expeditionary Forces in three locations.

Figure 24: Group 1 SUAS Programs

Figure 25: Group 2 SUAS Programs

The Group 2 systems shown in Figure 25 reflect AFSOC's continuing technology effort to provide advanced RPA to the warfighter through teaming with AFRL and OSD. AFSOC is pursuing air-launched lethal and non-lethal capabilities, including larger recoverable systems that could be deployed off MQ-1/9 or manned aircraft.

4.6.2   Small Unmanned Renewable Energy Long Endurance Vehicle

An AFRL Aerospace Systems Directorate (AFRL/RQ) IPT, working in collaboration with AFSOC, USSOCOM, and the Army Program Manager, Unmanned Aircraft Systems, recently completed the design of a Group 1 SUAS hybrid fuel cell power system. It offers a remarkable endurance capability (more than triple that of a conventional system), while meeting the ruggedization and reliability requirements of the warfighter for global operations. The overall goal of the Small Unmanned Renewable Energy Long Endurance Vehicle (SURGE-V) program is to demonstrate that hybrid fuel cell power systems are a viable alternative to meet the propulsion and control power needs of a fieldable hand-launched SUAS, while offering both the endurance and payload power only available to larger Group 2 systems. Although previous AFRL programs were successful in demonstrating the endurance benefit of fuel cell hybrid power systems for SUAS applications, they also demonstrated that a higher degree of integration must be achieved for the SUAS to be both rugged and reliable while operating in a global environment. The SURGE-V program was developed as a result of these lessons learned and is focused on incorporating "real-world" environmental and operational factors into the design of the long-endurance power system. A fully integrated SURGE-V system is planned to be ruggedized and demonstrated by the end of calendar year 2013.

4.6.3   Advanced SUAS Hybrid Power System

AFRL/RQ is developing fuel cell hybrid power system technology to extend the endurance and power to support advanced payload capabilities for Group 1 expendable AL-SUAS applications. As the Air Force lead command for SUAS, AFSOC intends to use this capability to extend the range of onboard sensors, see below the weather, track multiple targets, increase target acquisition accuracy, and provide direct support to ground teams. This power system technology, leveraging the SURGE-V program, is expected to offer at least 10 times an increase in flight endurance, despite operating with additional constraints, such as extended stowage, remote launch from a tube with little to no initial airflow, and operation over broad temperature and altitude ranges. Flight demonstrations for the persistent AL-SUAS technology are anticipated in FY15.

4.7   Efficient Engines/Alternate Power

The rapid development and deployment of unmanned systems has resulted in a corresponding increased demand for more efficient and logistically supportable sources for propulsion and power. In addition to improving system effectiveness, these improvements have the potential to significantly reduce life-cycle costs. Though much research is underway, there are many challenges that remain. Continued investment and prototyping are critical to future platform capabilities.

4.7.1   Highly Energy-Efficient Turbine Engine

Highly Energy-Efficient Turbine Engine (HEETE) will demonstrate engine technologies that enable fuel-efficient, subsonic propulsion to support future extreme endurance and range requirements, including embedded engines incorporating complex inlets and exhausts. HEETE has two challenges relative to Group 5 UAS: embedding a high-bypass engine internal to the aircraft fuselage and delivering large amounts of electrical power regardless of throttle or flight condition.

The HEETE design provides very small, high-powered cores to enable high bypass within the diameter constraints of an internally packaged engine, and the propulsive efficiency is provided by highly efficient fans designed with the distortion tolerance needed to run behind complex inlets.

The HEETE cores run at pressure ratios greater than 2.3 times the current state-of-the-art systems. Such ratios provide high thermal efficiency, and when combined with integrated multispool power extraction, they enable greater tolerance of auxiliary power at high altitudes and with long endurance.

4.7.2   Efficient Small-Scale Propulsion

Efficient Small-Scale Propulsion (ESSP) demonstrates small-scale engine technologies that enable fuel-efficient, subsonic propulsion to support current and future endurance/range requirements for Group 2 through 5 UAS. ESSP increases capability through versatile pervasive turbine and internal combustion engine technologies applicable to RPA, power generators and cruise missiles.

ESSP turbine engines will provide increased speed, range, survivability, and VTOL capability. ESSP internal combustion engines, such as the nutating engine, are lightweight and fuel efficient, having the power density and scalability for application in multiple UAS groups.

4.7.3   Alternative Fuels

Enhancement of RPA fuels for improved performance and mission capability has successfully demonstrated conversion from AvGas to JP-8, which has, in turn, relaxed critical and sensitive fuel performance requirements. This will result in significant fuel logistics savings while retaining or even improving RPA operability and reliability, compared to the baseline system.

4.7.4   X-51A Scramjet Engine

Hypersonic Technology's liquid hydrocarbon X-51A scramjet engine will enable the ability to defeat time-sensitive targets in an A2/AD environment from standoff ranges. At hypersonic speeds (600 nm in 10 minutes), the engine could also be employed on a future platform to provide time-critical ISR within 3,000 nm.

4.7.5   Technology Partnerships

The Air Force should continue to partner with National Aeronautics and Space Administration (NASA) and DARPA programs, which have potential to enhance mission capabilities such as persistence and fuel efficiency (e.g., NASA Environmentally Responsible Aviation Program, DARPA Falcon Project).

5.   PROCESS SYNCHRONIZATION AND IMPACT

5.1   Key DOD Corporate Processes

The Air Force operational capabilities requirements development process establishes the guidelines, policies, and procedures for defining, developing, documenting, validating, approving, and managing operational capability requirements supporting the Defense Acquisition Management Framework. This process integrates strategic planning, capabilities planning, early systems engineering, operational capability requirements development, acquisition, life-cycle management, and program and budget execution to effectively develop and field needed operational systems in a timely manner.

The Air Force operational capability requirements development process links directly to and complies with the Joint Capabilities Integration and Development System (JCIDS), QRC, and normal acquisition processes to deliver capabilities to warfighters and shape CFMP development. RPA stakeholders should leverage these DoD processes for RPA systems and sensor procurement.

Historical RPA procurement and fielding, for the most part, bypassed these key DoD processes. As a result, systems were procured through congressional inserts and sustained via OCO funding. During Air Force Corporate Structure deliberations, systems without documented requirement documents did not compete well for funding. Therefore, efforts are underway to retroactively document requirements for existing RPA systems while ensuring future requirements adhere to these key DoD processes.

5.2   JCIDS Process and Capabilities-Based Assessments

RPA stakeholders should consider desired capability effects and how they translate into warfighter requirements. This should include participation in the assessment of RPA contribution to Air Force Service Core Function (SCF) capabilities. The AFROC and JROC should be presented with a full picture of current and evolving capabilities to make informed decisions, including prioritization of identified capability gaps and requirements that satisfy those capability gaps.

CBAs are an integral part of the capabilities-based planning process, and their findings form the analytic basis for operational capability requirements development. The CBA defines the capability required and any capability gaps/shortfalls identified during the conduct of the assessment. Key components in a CBA consist of analysis of what the warfighter requires across all functional areas to accomplish the mission; a gap analysis of the capability needs against any existing or planned systems to identify associated gaps/shortfalls or redundancies; and recommendations on whether the gaps/shortfalls can be addressed by non-materiel means, materiel means, or both. The results of the CBAs are then documented in one of two documents: a Joint DOTMLPF Change Recommendation or an ICD. The designated AF/A5R division chief will work with the respective CFLI representatives and the sponsor to implement CBA results to the respective SCFs.

5.3   Requirements Development and JCIDS Integration

Strategic vision in this RPA Vector and subsequent versions should guide timely development and fielding of affordable and sustainable operational systems needed by the CCDR. JCIDS integrates with the acquisition and the planning, programming, budgeting, and execution processes to support improvements to existing warfighting capabilities and facilitates development of new warfighting systems. The blended process then validates warfighting capability needs while considering the full range of materiel and non-materiel solutions. Operational capabilities must be defined within the "art of the possible" and grounded within real-world constraints of time, technology, and affordability.

5.4   Air Force Requirements Oversight Council

The AFROC is an instrument of the Chief of Staff of the U.S. Air Force (CSAF), established to review, validate, and approve Air Force operational capability requirements. It assures Air Force documentation is prepared in accordance with appropriate Air Force and Joint Staff guidance, complies with established standards, and accurately articulates valid Air Force operational capability requirements before and during the acquisition processes. The AFROC's purview includes JCIDS, Urgent Operational Needs (UON)/Joint Urgent Operational Needs (JUON)/QRC, Resource Management Decision, Joint Capability Technology Demonstration (JCTD), S&T, and task forces. In addition, the AFROC reviews JCTD candidates and select S&T efforts to validate that they are addressing Air Force capability needs and developmental planning.

The AFROC will provide advocacy for approved/validated operational requirements resulting from capability gaps and shortfalls identified by the capabilities-based planning process. It is responsible for the standardization and quality of Air Force operational capability requirements processes and products. The AFROC also coordinates with other HAF directorates to resolve requirements, acquisition, and programmatic issues for all programs, including special access programs.

5.5   Quick Reaction Capability Process

RPA stakeholders and lead commands should leverage the Air Force's QRC process as delineated in AFI 63-114 to meet JUON, UON, or CSAF-directed activities. QRC programs provide limited materiel solutions for urgent warfighting needs and are resourced as high Air Force priorities. The Milestone Decision Authority (MDA) shall convene a capability transition review (CTR) no later than 180 calendar days following initial fielding. At the CTR, the MDA and lead command review the fielded capability's assessed suitability and effectiveness and formally document decisions regarding further development or disposition.

5.6   Core Function Master Plans

SCFs are functional areas that delineate the appropriate and assigned core duties, missions, and tasks of the Air Force as an organization. CFLIs are assigned responsibility for each of these SCF functional areas. They act as the principal integrators for their assigned SCF and the corresponding CFMP. CFMPs link Air Force strategic guidance to Air Force programming guidance by shaping the operational and resourcing health of each SCF across the spectrum and/or domain.

As an example, the GIISR CFMP captures Air Force strategic vision and enables cross-domain synchronization and integration of planning and operations of all ISR assets, sensors, PED, and analysis and production capabilities across the globe maximizing joint force battlespace awareness.

As a cross-domain asset, RPA capabilities should be considered by other CFLI across the full spectrum of SCFs, and this RPA Vector is intended as a reference to guide development of other Air Force CFMP.

5.7   Mission Integration with Core Function Master Plan

This document is intended to inform development of Air Force CFMPs and to look across multiple SCF portfolios. This document is intended to replace the CSAF- and SECAF-approved *Air Force UAS Flight Plan* 2009–2047, which pre-dated the CFLI construct. It will be reviewed and updated within 2 years by HAF functionals in coordination with CFLI staffs as required.

The CFMP forms a common framework linking strategic planning and programming to improve what the Air Force brings to the joint fight. In support of this, the CFLIs provide agile leadership to help the Air Force achieve the strategic and operational objectives of the National Defense Strategy with projected resources at the lowest possible overall risk. The 13 SCFs align with the Air Force's specific military service functions listed in DoD Directive 5100.01. It is the responsibility of lead MAJCOMs to establish enabling concepts, draft requirements, and accomplish all aspects of the organize/train/equip mission. The lead MAJCOM for Group 4 and 5 ISR/Strike RPA is ACC. The lead MAJCOM for airlift and AR RPA is Air Mobility Command. The lead MAJCOM for SUAS is AFSOC. The lead MAJCOM for RPA communications architecture is AFSPC.

5.8   RPA and SUAS Acquisition Recommendations

Currently RPA acquisition is stove-piped by weapon systems. There are a number of aspects, for example, that are common to Group 4 and 5 RPA that would benefit from common coordinated approaches. Some of these common efforts include data links, SAA systems, and standard interfaces. AFMC, in partnership with the acquisition community, will focus on full institutional integration of RPA with the Air Force sustainment and test and evaluation communities to ensure successful Air Force RPA development. The goal is to foster appropriate joint RPA acquisition with emphasis on innovation, rapid acquisition, and fielding. Ideally, the Air Force will be recognized as an RPA acquisition center of excellence, delivering joint RPA capabilities with best practices that can be exported across the DoD.

The Air Force must employ leading-edge technologies to the development of NextGen RPA capabilities and establish better communication with stakeholders and industry. To this end, the Air Force must determine the best method of applying any evolutionary requirements identified in the RPA Vector. The Air Force will apply the most current CJCSI 3170 and DoDI 5000.02 guidance for RPA acquisition, while also adopting the acquisition lessons learned and formalizing those lessons as part of Develop and Sustain Warfighting Systems efforts. To incentivize fair and open competition in the process, the Air Force will work with DoD and industry to establish common standards.

In the case of SUAS, one of the major challenges is that the systems evolve very rapidly as this technology grows with emerging lightweight payloads, improved C2 technology, and advances in manufacturing and miniaturization technologies. Often, the traditional requirements and acquisition cycle may not permit a capability to be delivered before it is already obsolete. As such, many of these systems have been acquired through rapid fielding programs. Another challenge with SUAS is that the technology refresh rate is comparable to computers or cell phones. The Air Force must consider an iterative spiral development program structure that allows for fielding of cutting edge technologies to meet emerging warfighter needs well inside of the typical acquisition timeline.

A critical technique to effectively manage and sustain RPA systems will be for the Air Force to procure the appropriate level of data rights and contract sustainment support to maintain a degree of organic systems engineering authority and system integrator responsibilities. Currently, the Air Force does not own the data rights for MQ-1, MQ-9, RQ-4, or any existing Group 1 UAS. This management action is essential for future systems to retain the ability to define and oversee the details of the integrated RPA environment. This would begin with the stated requirements and then build on the MAJCOM-developed concepts of employment to aid in defining the optimum suite of technologies that would best fill the capability requirement. DoDI 5000.02 prescribes the specific requirements for RDT&E. As technologies are developed, they will be demonstrated in an operationally relevant increment so they can be further matured, while the force provider continues to refine the requirement and facilitates the synchronization of all other actions. This requires a test and evaluation strategy (TES) for RPA platforms and payloads to address the unique aspects of each system and how it will integrate as a SoS. In the process, the SOA interface standards would be refined. ASC would apply this to optimize the suite of technologies for the MAJCOM-defined SoS architecture. The TES for RPA would address other unique challenges of testing FoS platforms and payloads that include selecting the responsible test organization (RTO) for developmental testing, possibly considering the contractor as the RTO, addressing contractor proprietary information, testing airspace access under current FAA rules, range safety, data telemetry and incremental development of capabilities. AFMC will determine resources needed for these actions, which may include increased funding and manpower.

5.9   Communications and Public Affairs

Effective communication is an operational imperative to gain and maintain credibility while increasing the understanding of and support for RPA operations. A command-supported, proactive communication program, hinged on communicating timely, accurate, and truthful information, is vital to supporting the Air Force's RPA mission across the ROMO and showcasing Air Force capability to global audiences. As such, capitalizing on outreach opportunities is integral to mission success and directly supports the DoD policy of "maximum disclosure with minimal delay" regarding coverage of military activities, including people, assets, and operations.

Air Force public affairs practitioners actively seek opportunities to integrate and synchronize communication to inform key audiences about RPA forces, capabilities, and requirements in support of the joint warfighter and the Air Force's mission, people and future. Public affairs professionals are charged with developing innovative methods for enabling and synchronizing enterprisewide communications and ensuring these fall within established Air Force public affairs guidelines and are appropriately coordinated with MAJCOMs and HAF. All public affairs activities are carried out in accordance with AFI 35-101 (Public Affairs Policies and Procedures), AFDD 2-5.3 (Public Affairs Operations), and the Air Force ISR Public Affairs Guidance across the information domain, including print, television, radio, web-based media, and speaking opportunities at venues including conferences, tradeshows, and community events. Communication campaign strategies are executed at the senior levels of government by appropriate Air Force leadership to enhance leaders' and lawmakers' understanding of RPA current and future roles.

Communication plans and public affairs guidance on the RPA force, capabilities and requirements have been developed to provide public affairs practitioners and leadership with strategic guidance regarding activities related to RPA operations. These products are living documents that are updated as information changes. Current public affairs activities include identifying outreach efforts to present the Air Force's RPA vision to DoD, other government users, academia and industry. This is accomplished through strategic participation at key conferences, site visits to service RPA facilities, and the development of collaborative relationships.

APPENDIX A   REFERENCES

1. *Air Force Association speech*, Donley, Michael B., Secretary of the Air Force, Orlando FL, 26 February 2009.

2. *Air Force Association's 2011 Air and Space Conference and Technology Exposition speech*, Donley, Michael B., Secretary of the Air Force, National Harbor MD, 19 September 2011.

3. *An Air Force Strategic Vision for 2020-2030,* Shaud, John A. (General, USAF, Ret.) and Lowther, Adam B., Maxwell AFB, AL: Air University Press, 2010.

4. *Airspace Integration Plan for Unmanned Aviation*, OSD, 2011.

5. *The Continued Growth of Unmanned Aircraft Systems Memo*, Secretary of the Air Force, 24 June 2011

6. *CSAF Vector 2011,* Schwartz, Norton A. (General, USAF), CSAF, 4 July 2011.

7. *Existing Joint Capability Integration and Development System Requirements*, A5RI, 2008.

8. *Focused Long-Term Challenges Overview (and current Ongoing Technology Efforts*, Air Force Research Lab (AFRL/XP), 13 October 2008.

9. *How might people interact with agents?,* Norman Donald A., Published in *Software Agents*, Bradshaw J. M. Ed., Cambridge, MA: The AAAI Press/The MIT Press, 1997, pp. 49–55.

10. *Joint Doctrine Note 2/11: UK Approach to Unmanned Systems,* Ministry of Defense, The Development, Concepts and Doctrine Centre, 2011.

11. *National Information Assurance Glossary. CNSS Instruction No. 4009*, Committee on National Security Systems, 2010.

12. *The National Military Strategy of the United States of America 2011*, Chairman, Joint Chiefs of Staff, 8 February 2011.

13. *NATO Industrial Advisory Group, Study Group 75, Annex C—Autonomous Operations*, 2004.

14. *Reliability, Availability and Maintainability Policy Memo*, SAF/AQ, 28 August 2008.

15. *Report on Technology Horizons: A Vision for Air Force Science and Technology During 2010–2030 Volume 1,* United States Air Force Chief Scientist, 15 May 2010.

16. *Schwartz outlines possible future changes,* Fontaine, Scott, Air Force Times, 30 August 2010.

17. *Sustaining U.S. Global Leadership: Priorities for 21st Century Defense,* OSD, January 2012.

18. *Unmanned Systems Integrated Roadmap FY2012–2037,* OSD, 2012.

19. *U.S. Navy Strategic Vision Briefing*, OPNAV Aviation Manpower/Training Branch (N882), 18 December 2008.

20. *Unmanned Aircraft Systems Roadmap 2010–2035*, United States Army, 2010.

21. *Air Force Aerial Layer Network Enabling Concepts and Roadmap,* Air Force, 2012.

APPENDIX B ACRONYMS

Acronym	Definition
A2/AD	Anti-Access / Area Denial
AAR	Automated Air Refueling
ABSAA	Airborne Sense and Avoid
ACC	Air Combat Command
ACES–HY	Airborne Cueing and Exploitation System Hyperspectral
ADS-B	Automatic Dependent Surveillance-Broadcast
AEHF	Advanced Extremely High-Frequency
AES	Advanced Encryption Standard
AESA	Active Electronically Scanned Array
AFDD	Air Force Doctrine Document
AFI	Air Force Instruction
AFMC	Air Force Materiel Command
AFR	Air Force Reserve
AFRL	Air Force Research Laboratory
AFROC	Air Force Requirement Oversight Council
AFROCM	Air Force Requirement Oversight Council Memorandum
AFSC	Air Force Specialty Code
AFSMO	Air Force Spectrum Management Office
AFSOC	Air Force Special Operations Command
AFSPC	Air Force Space Command
AI	Artificial Intelligence
AL-SUAS	Air-launched Small Unmanned Aircraft System
ANG	Air National Guard
AoA	Analysis of Alternatives
AOR	Area of Responsibility
API	Application Programming Interface
AR	Air Refueling
ASC	Aeronautical Systems Center
ATAO	Autonomous Terminal Area and Ground Operations
ATC	Air Traffic Control
BACN	Battlefield Airborne Communications Node
BDA	Battle Damage Assessment
BE	Bandwidth Efficient
BLOS	Beyond Line-of-Sight
BMC2	Battle Management Command and Control
C2	Command and Control
C4ISR	Command, Control, Communications, Computers, Intelligence Reconnaissance, and Surveillance
CAMARO	Cooperative Automated Multi-aircraft RPA Operations
CAP	Combat Air Patrol
CAS	Close Air Support
CASEVAC	Casualty Evacuation
CBA	Capabilities-Based Assessment
CBRNE	Chemical, Biological, Radiological, Nuclear, and Explosive
CCDR	Combatant Commander
CDL	Common Data Link
CFLI	Core Function Lead Integrator

Acronym	Definition
CFMP	Core Function Master Plan
CID	Combat Identification
CJCS	Chairman of the Joint Chiefs of Staff
CJCSI	Chairman of Joint Chiefs of Staff Instruction
COA	Certificate of Authorization
COMSATCOM	Commercial Satellite Communications
CONOPS	Concept of Operations
CONUS	Continental United States
COP	Common Operational Picture
COTS	Commercial Off-The-Shelf
CQ-X	NextGen RPA Transport
CSAF	Chief of Staff of the U.S. Air Force
CSAR	Combat Search and Rescue
CSI	Common Systems Integration
DARE	Distribution Access Range Extension
DARPA	Defense Advanced Research Projects Agency
DCGS	Distributed Common Ground System
DDR	Dismount Detection Radar
DGS	Distributed Ground Station
DI2E	Defense Intelligence Information Enterprises
DIB	DCGS Integration Backbone
DIME	Dynamic Immersive Mission Environment
DMO	Distributed Mission Operations
DoD	Department of Defense
DoDI	Department of Defense Instruction
DOTMLPF	Doctrine, Organization, Training, Materiel, Leadership and education, Personnel, and Facilities
DSA	Dynamic Spectrum Access
DSCA	Defense Support to Civilian Authorities
DYNAMAN	Dynamic Maneuvering
EA	Electronic Attack
EO	Electro-optical
ESSR	Efficient Small-Scale Propulsion
EW	Electronic Warfare
FAA	Federal Aviation Administration
FAC-A	Forward Air Controller-Airborne
FLEX-IT	Flexible Levels of Execution Interface Technologies
FMV	Full Motion Video
FOPEN	Foliage Penetrating
FoS	Family of Systems
FY	Fiscal Year
GBSAA	Ground-Base Sense and Avoid
GCC	Geographic Combatant Command
GCS	Ground Control Station
GEOINT	Geospatial Intelligence
GIG	Global Information Grid
GIISR	Global Integrated Intelligence Reconnaissance and Surveillance
GIS	Geospatial Information System
GMTI	Ground Moving Target Indicator

Acronym	Definition
GPA	Global Precision Attack
GPS	Global Positioning System
HAF	Headquarters Air Force
HALE	High-Altitude Long-Endurance
HCB	High-Capacity Backbone
HD	High Definition
HEETE	Highly Energy-Efficient Turbine Engine
HMI	Human-Machine Interface
HSI	Human-Systems Integration
IADS	Integrated Air Defense Systems
ICAO	International Civil Aviation Organization
ICD	Initial Capabilities Document
IED	Improvised Explosive Device
INMARSAT	International Maritime Satellite
IOP	Interoperability Profile
IP	Internet Protocol
IPT	Integrated Process Team
IR	Infrared
ISR	Intelligence, Surveillance, and Reconnaissance
IW	Irregular Warfare
JALN	Joint Aerial Layer Network
JCD	Joint Capabilities Document
JCIDS	Joint Capabilities Integration and Development System
JCTD	Joint Capability Technology Demonstration
JFC	Joint Force Commander
JOAC	Joint Operational Access Concept
JP-8	Jet Propellant 8
JROC	Joint Requirements Oversight Council
JROCM	Joint Requirements Oversight Council Memorandum
JTAC	Joint Terminal Attack Controller
JUON	Joint Urgent Operational Needs
KCQ-AR	Future RPA for Air Refueling
LOC	Lines of Communication
LOS	Line-of-Sight
LPI/LPD	Low Probability of Intercept or Detection
LRE	Launch and Recovery Element
MAC	Multi-aircraft Control
MAJCOM	Major Command
MAM	Multi-aircraft Manager
MCE	Mission Control Element
MEDEVAC	Medical Evacuation
MIC	Mission Intelligence Coordinator
MILSATCOM	Military Satellite Communications
MP-RTIP	Multiplatform Radar Technology Insertion Program
MUM	Manned/Unmanned
NAS	National Airspace System
NATO	North Atlantic Treaty Organization
NCOE	Net-Centric Operational Environment
NET-T	Network-Tactical

Acronym	Definition
NextGen	Next-Generation
NSA	National Security Agency
OCO	Overseas Contingency Operations
OCONUS	Outside the Continental United States
OEF	Operation Enduring Freedom
OIF	Operation Iraqi Freedom
OODA	Observe, Orient, Decide, and Act
OPA	Optionally Piloted Aircraft
OSD	Office of the Secretary of Defense
PED	Processing, Exploitation, Dissemination
PIC	Pilot in Command
PMATS	Predator Mission Aircrew Training System
PNT	Position, Navigation and Timing
PoR	Program of Record
PR	Personnel Recovery
QRC	Quick Reaction Capability
R&D	Research and Development
R&R	Relief and reconstruction
RAIN	Remotely Accessible Internet Protocol Network
RDT&E	Research, Development, Test and Engineering
RF	Radio Frequency
ROCC	RPA Operations Coordination Center
ROMO	Range of Military Operations
ROVER	Remotely Operated Video Enhanced Receiver
RPA	Remotely Piloted Aircraft
RSO	Remote Split Operations
RVT	Remote Video Terminal
S&T	Science and Technology
SA	Situational Awareness
SAA	Sense and avoid
SAR	Synthetic Aperture Radar
SATCOM	Satellite Communications
SCAR	Strike Coordination and Reconnaissance
SCF	Service Core Function
SEAD	Suppression of Enemy Air Defense
SECDEF	Secretary of Defense
SIGINT	Signals Intelligence
SIRIS	Surveillance, Intelligence Reconnaissance, Information System
SOA	Service-Oriented Architecture
SOC	Squadron Operations Center
SOF	Special Operations Forces
SoS	Systems of Systems
SOWT	Special Operations Weather Team
SPO	System Program Office
STANAG	Standardization Agreement
STD-CDL	Standard Common Data Link
STUAS	Small Tactical Unmanned Aircraft System
SUAS	Small Unmanned Aircraft System
SUAS-O	Small Unmanned Aircraft System Operator

Acronym	Definition
SURGE-V	Small Unmanned Renewable Energy Long Endurance Vehicle
SV	Satellite Vehicles
SWaP	Size Weight and Power
TBM	Theater Ballistic Missile
TOC	Tactical Operations Center
TTP	Tactics, Techniques, Procedures
UA	Unmanned Aircraft
UAI	Universal Armament Interface
UAS	Unmanned Aircraft System
UCI	UAS C2 Initiative
UCS	UAS Control Segment
UDOP	User Defined Operating Picture
UHF	Ultra High Frequency
UON	Urgent Operational Needs
USIP	Universal System Interoperability Protocol
USSOCOM	United States Special Operations Command
VTOL	Vertical Takeoff and Landing
WAS	Wide Area Sensor
WGS	Wideband Global SATCOM
WRC	World Radio communication Conference

APPENDIX C DEFINITIONS

Family-of-Systems (FoS): UAS with similar operating characteristics and control interfaces that are typically, but not always, provided from a single manufacturer. FOS designation will be determined by the lead command.

Launch and Recovery Element (LRE): The element at the forward operating location that consists of aircraft, a launch and recovery GCS, and associated required communications equipment. The crews deployed to LREs are responsible for launching and recovering the aircraft.

Mission Control Element (MCE): The element composed of the GCSs, SOC, and communications equipment required for RSO and data distribution. The MCE serves as the core for mission coordination, fused intelligence, planning, and execution.

Operator: The individual monitoring and controlling a SUAS through issuance of command input to the aircraft. SUAS operators are typically enlisted Airman possessing the appropriate training, certifications, and ratings.

Pilot: The individual monitoring and controlling of an RPA through issuance of command input to the aircraft. RPA pilots are rated pilots and possess the appropriate training, certifications, and ratings.

Pilot in Command (PIC): The person who has final authority and responsibility for the operation and safety of flight, has been designated as pilot in command before or during the flight, and holds the appropriate category, class, and type rating, if appropriate, for the conduct of the flight. The responsibility and authority of the PIC as described by 14 Code of Federal Regulations Part 91.3, *Responsibility and Authority of the Pilot in Command*, apply to the remotely piloted aircraft PIC. The PIC position may rotate duties as necessary with equally qualified pilots. The individual designated as PIC may change during flight.

Remote Split Operations (RSO): The geographical separation of the launch and recovery "cockpit" and crew from the mission "cockpit" and crew. RSO enables the employment of the aircraft by the mission crew at a location other than where the aircraft are based (in some cases thousands of miles from the actual aircraft location). The team/location/equipment that launches and recovers the aircraft is referred to collectively as the LRE. The GCS where the mission is flown from is called the MCE.

Remotely Piloted Aircraft (RPA): The aircraft portion of the broader UAS classification. (Air Force definition) Currently, Group 4 and 5 UA are considered RPA and are flown by rated pilots. An RPA requires a pilot, sensor operator, GCS, associated manpower and support systems, and satellite communication links to perform its mission and intelligence integration. RPA engage in many of the same missions as manned aircraft, such as CAS, ISR, dynamic targeting, and air interdiction. Examples of RPA include the MQ-1 Predator, MQ-9 Reaper, and RQ-4 Global Hawk.

Sensor Operator: The person who controls the payloads or sensors onboard an RPA but does not command and control the aircraft. (Air Force definition)

Small Unmanned Aircraft System (SUAS): A Group 1-3 UA operated by a qualified SUAS-O who, while not a rated pilot, functions as the pilot-in-command of the SUAS. (Air Force definition) Examples of SUAS include the Raven, Puma, and ScanEagle.

Squadron Operations Center (SOC): The element serving as the core for communications and network components required to conduct RSO, mission data dissemination, and RPA C2. The SOC provides the core capability for RPA mission planning and SA, including fused intelligence, updates on weather, threats, targets that enable mission execution.

System of Systems (SoS): A set or arrangement of interdependent systems that are related or connected to provide a given capability that enables cost effective measures that increase capabilities by distributing weapon and sensor capabilities across a formation of aircraft. The loss of any part of the system significantly degrades the performance or capabilities of the whole. Individual vehicle capabilities and payloads can be tailored and scaled to mission needs. The avionics architecture and sensors on the aircraft must be capable of rapid changes of payload types and provide users and maintainers with plug-and-play capability.

Unmanned Aircraft (UA): An aircraft or balloon that does not carry a human operator and is capable of flight under remote control or semi-autonomous programming.

Unmanned Aircraft System (UAS): A system consisting of a control station, one or more unmanned aircraft, control and payload data links, and mission payloads, designed or modified not to carry a human pilot and to be operated through remote or self-contained semi-autonomous control.